PREHISTORIC AUSTRALASIA

VISIONS OF EVOLUTION AND EXTINCTION

MICHAEL ARCHER, SUZANNE J. HAND,
JOHN LONG, TREVOR H. WORTHY
AND PETER SCHOUTEN

CSIRO

PUBLISHING

A catalogue record for this book is available from the National Library of Australia.

ISBN: 9780643108059 (hbk)
ISBN: 9780643108066 (epdf)
ISBN: 9780643108073 (epub)

How to cite:
Archer M, Hand SJ, Long J, Worthy TH, Schouten P (2023) *Prehistoric Australasia: Visions of Evolution and Extinction.* CSIRO Publishing, Melbourne.

Published by:

CSIRO Publishing
36 Gardiner Road, Clayton VIC 3168
Private Bag 10, Clayton South VIC 3169
Australia

Telephone: +61 3 9545 8400
Email: publishing.sales@csiro.au
Website: www.publish.csiro.au
Sign up to our email alerts: publish.csiro.au/earlyalert

Front cover: A pair of Australian predatory theropods, *Australovenator wintonensis* (artwork by Peter Schouten)
Back cover: (top to bottom) The Shark-toothed Dolphin, an unnamed squalodontid; mini Marsupial Lion, *Lekaneleo roskellyae*, stalking the enigmatic marsupial *Yalkaparidon coheni*; the carnivorous macropod *Propleopus*, preying on an Australian Brush-turkey (artworks by Peter Schouten)
Endpapers: Tetrapod forelimbs (artwork by Peter Schouten)

Edited by Adrienne de Kretser, Righting Writing
Cover, text design and typeset by Cath Pirret Design
Printed in China by Leo Paper Products Ltd

CSIRO Publishing publishes and distributes scientific, technical and health science books, magazines and journals from Australia to a worldwide audience and conducts these activities autonomously from the research activities of the Commonwealth Scientific and Industrial Research Organisation (CSIRO). The views expressed in this publication are those of the author(s) and do not necessarily represent those of, and should not be attributed to, the publisher or CSIRO. The copyright owner shall not be liable for technical or other errors or omissions contained herein. The reader/user accepts all risks and responsibility for losses, damages, costs and other consequences resulting directly or indirectly from using this information.

CSIRO acknowledges the Traditional Owners of the lands that we live and work on across Australia and pays its respect to Elders past and present. CSIRO recognises that Aboriginal and Torres Strait Islander peoples have made and will continue to make extraordinary contributions to all aspects of Australian life including culture, economy and science. CSIRO is committed to reconciliation and demonstrating respect for Indigenous knowledge and science. The use of Western science in this publication should not be interpreted as diminishing the knowledge of plants, animals and environment from Indigenous ecological knowledge systems.

The paper this book is printed on is in accordance with the standards of the Forest Stewardship Council® and other controlled material. The FSC® promotes environmentally responsible, socially beneficial and economically viable management of the world's forests.

Sept23_R01

Contents

Introduction

PALAEONTOLOGISTS CAN NO MORE STOP DIGGING UP FOSSILS THAN STOP breathing – it's an obsession! Australasian palaeontologists are particularly driven to do this because the biological histories of Australia and its smaller neighbour, Zealandia, are the least understood of all the occupied continents. Hence, every new discovery can have a huge impact on what we previously understood about the once normal inhabitants of this very unusual continent. However, while discovery of a new tooth row or skull can send palaeontologists into fits of excitement, for the general public to share in the excitement of discovery, it often takes the palaeoartist's multifaceted skills to transform the scientists' discoveries into the next best thing to flesh and blood. To produce this book, *Prehistoric Australasia: Visions of Evolution and Extinction*, many palaeontologists have worked closely for years with palaeoartist Peter Schouten to produce a unique series of panoramas, windows in time that span the last 3.6 billion years of extraordinary life of Australasia and the south-west Pacific.

A long and complicated history of becoming different

Rudyard Kipling produced a *Just So* story about how the ancestral non-hopping kangaroo became unlike all other animals after commanding its creator to 'Make me different from all the other animals by five o'clock this afternoon!' The wish was granted, giving the kangaroo powerful hopping legs just in time for it to avoid becoming a Dingo's dinner. In reality, it took a fair bit longer than a day for kangaroos and Australia's other animals to become different. In fact, it took many millions of years for this area of the world to evolve the biggest dinosaurs in the world, gigantic dromornithid birds, bizarre thingodontans, marsupial lions, spine-bristled echidnas, wombats, koalas, kangaroos, horned turtles and even gum trees. How they have changed over time is a distinctly Australasian *Just So* story, and the focus of this book.

For much of its geological history, Australia's creatures rather closely resembled those of the other continents. From at least as long ago as 3.6 billion years, the part of Earth's crust and associated inland seas that were to become Australia were often physically connected to the other continents. As a result, many kinds of life – from cyanobacteria to trilobites, lungfish, sauropod dinosaurs, megaraptors and cynodonts that thrived in Australia – belonged to groups that spanned the united lands and waterways of Earth.

About 335 million years ago, all of the continents were joined and together they formed the supercontinent Pangaea. But because continental scrums of this type have always had limited durations, Pangaea began to tear itself apart around

175 million years ago. In the crustal destruction that followed, the island continent of Australia was eventually born, separating from Antarctica first along its western edge about 100 million years ago and then finally from its eastern edge sometime between 50 million and 38 million years ago. With the severing of the continent's last ties to what was left of Pangaea, the now isolated creatures of Australia began to evolve in directions unlike those anywhere else on Earth. The origin of New Zealand is closely tied to that of Australia; until 80–60 million years ago both were part of Gondwana. Then Zealandia, the India-sized continent of which New Zealand and New Caledonia are the largest emergent parts, unzipped from Australia's eastern flank, and the world's smallest continent began its own unique 80 million years history of isolation

As this antipodean ark set out on its steady journey northward through the Indian Ocean at the breakneck pace of around 7 cm per year, global climatic cycles, alternating between greenhouse and icehouse conditions, tested, prodded and transformed the continent's biota. This tapestry of changes accumulated because Australia was no longer exchanging its land-based creatures with those of the other continents. For the last 50 million years, elephants, camels, tapirs, cats, sloths, anteaters, primates, bears, horses and many other groups were able to move back and forth between the frequently connected lands of Asia, Africa, Europe and North and South America, but none of these globally widespread groups could reach the island continents of Australia and Zealandia.

The small maps associated with the panoramas that follow provide a snapshot of how Australia's (and Zealandia's) geographic position has been changing over time.

An independent laboratory churning out evolutionary innovations

While it is true that many of this region's living and recently extinct creatures represent outcomes of independent global experiments in evolutionary history, it has also been the case that this portion of the ancient world is the first known to have developed many of the most interesting steps in the evolutionary history of life as a whole. In fact, the oldest known solid objects of any kind found on planet Earth are 4.45 billion years old zircon crystals from the Jack Hills area of Western Australia.

In terms of explosive moments in the history of life, this region of the world has produced some of the most important. Here are 35 of the region's notable world fossil records.

1. The first undoubted evidence for life forms on planet Earth (at least 3.5 billion years ago; however, there are controversial contenders for this title in Greenland at 3.7 billion years ago; p. 18).

2. The oldest undoubted cells (in the ~3.5 billion years old Apex Chert Formation from Western Australia).

3. Seriously weird early organisms at 3.4 billion years ago (from the Strelley Pool Formation of Western Australia).

4. Among the oldest, if not the oldest, undoubted eukaryotic cells and hence routine practice of sexual reproduction (1.5 billion years old *Tappania* from the Roper Group of the Northern Territory).

5. Among the oldest evidence for Earth's earliest animals (e.g. *Myxomitodes stirlingensis*, the so-called mucous monsters at 1.2 billion years ago from the Stirling Ranges in Western Australia; or 650 million years old sponges from the Flinders Ranges in South Australia).

6. Among the best-known early communities of animals (570 million years old Ediacaran vendozoans from South Australia; p. 20).

7. The earliest perfectly formed eyes in predatory animals (525 million years old anomalocarids from the Emu Bay Shale, South Australia; p. 22).

8. Possibly the world's first known fossil bone (495 million years old bone bits from the Gola Beds of Queensland, although this is controversial – it could be arthropod cuticle).

9. Evidence for the world's first forests (~400 million years old root systems from the Wee Jasper Formation of New South Wales).

10. Among if not the world's first fish with bone (458 million years old *Arandaspis* from the Stairway Sandstone, Northern Territory; p. 24).

11. Possibly the world's first shark, *Tantalepis* (458 million years old, known from placoid scales identical to those seen in modern sharks, from the Stairway Sandstone, Northern Territory).

12. Among the world's earliest land plant communities (e.g. the 427 million years old *Baragwanathia* Flora from Victoria; p. 26).

13. Among the first evidence of sexual intercourse (e.g. 385 million years old placoderm fish from Gogo, Western Australia as well as Scotland; p. 36).

14. First evidence of live birth in a vertebrate (380 million years old *Materpiscus* from Gogo, Western Australia; p. 36).

15. Among the world's first fully terrestrial, four-footed animals (340 million years old *Ossinodus* from Ducabrook, Queensland; p. 44).

16. The world's biggest dinosaurs (130 million years old individual sauropod tracks 1.75 m in length on the coast north of Broome, Western Australia; p. 62).

17. The longest surviving 'labyrinthodont' amphibian (*Koolasuchus* from 120 million years old sediments near Inverloch, Victoria).

18. The first of the world's 'modern crocodiles' (eusuchians) (*Isisfordia* from Winton in Queensland and Lightning Ridge, New South Wales at 100 million years old; p. 80).

19. Among if not the world's first megaraptors (100 million years old 'Lightning Claw' from Lightning Ridge, New South Wales; p. 84).

20. The world's first egg-laying monotremes (e.g. the 113 million years old *Teinolophos* from Flat Rocks, Victoria).

21. The world's oldest passeriform (song) birds (55 million years old limb bones from Murgon, Queensland, are at least 20 million years older than songbirds anywhere else; p. 96).

22. Among if not the world's oldest bat fossils (55 million years old *Australonycteris* from Murgon, Queensland; p. 98).

23. The world's only known at least partially arboreal 'drop crocs' (23–15 million years old possibly arboreal mekosuchine crocodiles from Riversleigh, Queensland; p. 138).

24. The world's only 'drop bears' (actually *Nimbadon lavarackorum*, arboreal sloth-like marsupials at 15 million years old from Riversleigh, Queensland; p. 140).

25. Among the world's oldest known sperm cells which also have nuclei preserved inside them (17 million years old ostracod cells from Riversleigh, Queensland).

26. The world's biggest birds (e.g. 8 million years old flightless dromornithid species of *Dromornis* from Alcoota in the Northern Territory and the extraordinary radiation of giant flightless ratites, the moa, from New Zealand; p. 146).

27. The world's only known 'thingodontans', a whole order of mammals unique to Australia (23–15 million years old *Yalkaparidon* from Riversleigh, Queensland; p. 126).

28. The world's most specialised mammalian carnivores (24 million–35 000 years old marsupial lions from many areas of Australia; pp. 124, 126, 148, 154, 170, 174, 176).

29. The world's largest lizard as well as the largest venomous animal known (2 million–500 000 years old *Varanus priscus* from various regions of Australia; p. 172).

30. The world's largest marsupial (~2800 kg) from any continent or time, *Diprotodon optatum* (known from most of Pleistocene Australia; pp. 156, 162, 176, 190, 196).

31. The world's tallest trees of any kind (Mountain Ash, *Eucalyptus regnans*, individuals of which reach 150 m in height).

32. Possibly the longest-living organisms on planet Earth (e.g. King's Holly, *Lomatia tasmanicum*, individuals of which have been assessed to be up to 135 000 years old).

33. The world's biggest parrot (*Heracles*, a 1 m tall 7 kg giant from the 19–16 million years old St Bathans Fauna in New Zealand, nicknamed 'Squawkzilla'; p. 136).

34. The world's largest eagle anytime, anywhere (Haast's Eagle, Quaternary of New Zealand; p. 220).

35. The most massive galliform bird (distantly related to chickens) ever to have evolved (*Sylviornis* from Pleistocene deposits in New Caledonia; p. 216).

Given these and other 'Guinness Book of Life's Extreme Creatures' – many of which are illustrated in the panoramas that follow – this corner of the world has clearly punched well above its weight in terms of evolving extraordinary new forms of life. Many of them went on to have huge impacts on the rest of the world.

Putting the dead to work to save the living

There is growing awareness that fossils may help us save some of today's critically endangered species. We realise that we need to learn from the past to understand the present, in order to devise better conservation strategies for the future. All around the world, as climate change begins to destabilise habitats, conservation-minded palaeontologists are exploring the fossil record for clues about where threatened species might be able to be translocated to survive these changes. Often, this reveals options that might never have occurred to ecologists trying to solve the same problem. In most cases, clues are coming from Pleistocene records. This is why the critically endangered 'alpine' Takahe in New Zealand, the largest land rail in the world, was translocated out of the alpine zone to coastal islands where it had thrived during the Pleistocene – a translocation that has worked.

As this book has been coming together, some of us as palaeontologists are reaching even further back into Australia's fossil record for palaeontological clues about how we might save the critically endangered Mountain Pygmy-possum *Burramys parvus*. It was first named in the 1890s as a fossil recovered from a cave deposit in New South Wales before being found alive in 1966 in a ski lodge on Mount Hotham, Victoria. Ironically, this little possum is now threatened with becoming genuinely extinct as its alpine habitat is changed by climate heating. But rather than accept that this possum must become extinct, several of us have pointed out that it is the most recent member of a long chain of species in the same genus spanning at least the last 25 million years. And in every case known – *B. wakefieldi* from the 24 million years old Ngama Local Fauna of Lake Palankarinna in central Australia, *B. brutyi* from the 24–15 million years old faunal assemblages of Riversleigh in north-western Queensland, *B. triradiatus* from the 4.3 million years old Hamilton Local Fauna of western Victoria – these species have always lived in lowland wet forest habitats (p. 138). We suspect that at some time during the Pleistocene they followed shifting wet forest vegetation up the slopes of the Great Dividing Range then became stranded in the alpine zone when climates changed again. This is not an uncommon situation for endangered animals and plants around the world – they now occupy only marginal parts of their much larger former ranges. Based on this understanding, the Burramys Project (Fig. 1) involves construction of a lowland breeding facility in New South Wales, from where surplus individuals can be trial released into more stable cool temperate lowland wet forest habitats and monitored. If successful, similar translocation projects may be developed based on our growing knowledge about the fossil record.

A similar argument about using the fossil record to save other critically endangered species could be explored in relation to the Western Swamp Tortoise *Pseudemydura umbrina* in swamps along the western coast of Western Australia. With climate change and increasingly severe droughts, this last survivor of a highly distinctive group of turtles is headed for extinction. However, a species that is very closely related, if not the same, occurred in the middle Miocene rainforest pools in Riversleigh 15 million years ago (p. 138). Before the habitat of the living species disappears altogether, wouldn't it make sense to trial releasing a small population

A. Fossil record 25-0 Ma

25 million: Tirari Desert, S.A.

Burramys wakefieldi

Lowland wet forest

24-15 million: Riversleigh, Qld

Burramys brutyi

Lowland wet forest

4.3 million: Hamilton, Victoria

Burramys triradiatus

Lowland wet forest

B. The plan

Today: the 'Alps'

Burramys parvus

1. Live capture small number Mt Kosciuszko

2. Breed in Secret Creek Sanctuary in Lithgow, NSW

3. Research & acclimatise to release area resources

4. Release into lowland wet forest and monitor

5. Goal: Future secure in the habitat where their ancestors thrived.

Fig. 1. (A) Earlier species of *Burramys* are known from fossil deposits accumulated in cool temperate lowland wet forest environments. Reconstruction of late Oligocene habitat of *B. wakefieldi* in northern South Australia (J. Reece). Reconstruction of Miocene habitat of *B. brutyi* in the Riversleigh region of north-western Queensland (D. Dunphy). Modern wet forest similar to early Pliocene habitat of *B. triradiatus* in north-west Victoria (M. Archer). **(B)** The key steps involved in the conservation introduction proposal. *Burramys parvus* in alpine region (M. Archer). *B. parvus* in hands (H. Bates). Secret Creek facilities (M. Archer). Rainforest release site (M. Archer). Replicated small images representing *Burramys* possums are based on a photograph by Joel Sartore (ref. 15).

into cool temperate rainforest pools in eastern Australia that lack other turtles? Eventually, as conditions in Western Australia worsen, this option may become the only one.

As knowledge about the fossil record increases, it is also providing a new tool to independently assess the degree of concern we should feel about many of the endangered animals of Australia and its close neighbours. Modern ecological studies cannot provide a deep-time perspective about how lineages have fared through time and into the present. Hence, it is sometimes hard to assess whether these lineages have been declining, increasing or staying the same through time to bring them to the situation we find them in today. To make decisions about how to best use the limited funds available for conservation programs, it would be helpful to know which seemingly endangered species should get priority attention. For example, the long-term perspective about lineage change in Platypuses suggests that for at least the last 63 million years they have been declining in diversity (one species now), geographic distribution (just the rivers systems of eastern Australia) and morphological resilience (now being toothless and hyperspecialised). Palaeontology teaches us that when the last members of any lineage exhibit declines of this kind, even if they seem robust in their current situation they are actually in increasing danger of going extinct. A similar long-term decline in diversity and geographic area among Thylacines ('Tasmanian Tigers') – unbeknownst to the early colonists in Tasmania, many of whom persecuted them with the encouragement of the Tasmanian government of the time – ultimately led to their extinction. In contrast, although there has been a loss of many kinds of koalas over the last 25 million years and its numbers have recently declined largely because of habitat destruction, the modern Koala is far more abundant today in its forest habitat than any other koala species has ever been in the past. On this basis, palaeontologists, in contrast to some conservationists, would suggest that the Platypus is in far greater danger of extinction than the Koala.

Must all these extinct species stay extinct forever?

Most of us tend to think of the fossil record as a collection of fascinating curiosities, extinct creatures whose time in the sun has come and gone. Increasingly, however, there is interest in rethinking the 'gone' bit and it has given rise to the controversial new science of deExtinction. This involves questioning the once unarguable assumption that extinction is forever. Organisations like Revive & Restore and Geneticrescue have been stimulating scientists to think about the possibility of bringing extinct species back to life. While the introduction to *Jurassic Park* explains that the book was written to scare us into avoiding any temptation to use ancient DNA to resurrect prehistoric animals, both the book and the movie franchise that followed had exactly the opposite effect: they created intense interest in the possibility of carrying out such procedures with other extinct species.

Organisms as diverse as Siberian Mammoths, European Aurochs, North American Passenger Pigeons, Spain's Bucardos, Australia's Southern Gastric-brooding Frog and many others are now subjects of active research projects to challenge the assumption that extinction is forever. For some, such as the Mammoth Project and the Passenger Pigeon Project, this involves recovering ancient DNA from fossils and using it to transform the DNA of living relatives to produce new hybrid creatures that are physically and ecologically similar to the ones that went extinct. For others, such as the Auroch Project, research involves back-breeding to identify and resuscitate ancient genes that restore features of ancestral populations. For still others, such as the Lazarus Project focused on the Southern Gastric-brooding Frog (Fig. 2), study involves recovering whole nuclei with intact DNA from specimens of extinct species and introducing these into the enucleated egg cells of living relatives with the goal of bringing the whole of the extinct creature back to life. Even the possibility of revitalising lost dinosaurs is being explored through manipulation of the genome of the one group of dinosaurs that didn't go extinct – birds. A great deal of careful thought about risks is going into these projects. We need to assess, for example, whether the revived extinct species could be safely as well as usefully put back into ecosystems.

If deExtinction does become a reality, what other extinct Australasian species would we like to have back in the world? Certainly, the Thylacine would be a prime

Fig. 2. *Rheobatrachus silus*, the now extinct Southern Gastric-brooding Frog, shown in its original Queensland rainforest habitat. It is regurgitating its young which, remarkably, developed entirely in the adult's gut. This frog is now the focus of a deExtinction effort, the Lazarus Project, whose aim is to use nuclei from frozen tissues to bring this extraordinary, uniquely Australian species back to life.

candidate. We began recovering ancient DNA from Thylacines as early as 2000, with the ultimate goal of resurrection, and the whole mitochondrial and nuclear genomes have now been published – the 'recipe' for revitalised Thylacines. It has already been demonstrated that when some of the ancient DNA recovered from pickled Thylacine specimens is spliced into the genomes of a mouse, it works – it produces Thylacine tissues as part of hybrid embryos. Hopefully it will be only a matter of time before Thylacines are able to be returned to natural habitats in Tasmania and once again become Australia's King of Beasts. If we're successful in resurrecting Thylacines, would we consider the possibility of using ancient DNA, if it can be recovered, to try to resurrect the Thylacoleo, the magnificent lioness-sized marsupial and world's most specialised mammalian carnivore (p. 170)? How much more exciting camping trips into outback Australia would become! Projects of this kind are still a bit beyond our reach but, considering the pace at which synthetic biology is progressing, we could well have at least Thylacines back in the world by 2050.

Extinct animals are not really extinct – just paused in time

MIKE ARCHER

BUT WHAT DOES IT MEAN when we say something has become 'extinct'? To palaeontologists, whose grasp of reality focuses on the fourth dimension (time) as well as the other three (length, width and height), the idea of extinction can be complex. In the time dimension, every living cell in every organism that has ever lived on Earth is in physical contact with every other cell that has ever existed – there are challenges here to common perceptions about the nature of life. In this world view there are no gaps, and there never have been, between any of the species that have ever existed since life began on planet Earth perhaps 3.7 billion years ago. All life is a *single* time-travelling shape-changing organism, like a gigantic amoeba moving through space and time. For want of a better term, I have called it the 'Bioblob'. Among other insights provided by this view of the actual nature of life as a single time-travelling organism is the realisation that death cannot occur – it's an illusion of limited vision. Although we can understand the time dimension, we can't actually 'see' it with our eyes in the same way we see length, width and height. Hence, we perceive 'gaps' between individuals and species now and back through time despite the fact that, back along that dimension, all those parts of the Bioblob are physically interconnected parts of a single organism. And if death cannot occur, neither can extinction, at least not in the way it is conventionally conceived as an end to being alive. An 'extinct' part of the Bioblob is just a part of the organism that is not currently moving forward or shapeshifting in space and time. But despite its pause in transformation, like the nose on your face as you walk along, it's still part of the single living organism to which it belongs.

Finally, considering the Bioblob concept, if there has only been one organism, ever, on planet Earth, what should we think of fossils given that palaeontologists treat these as parts of extinct species? In actuality, fossils are the 'dandruff' and discarded fingernails of the Bioblob, shed along the way as it moves through time and space, transforming into new forms as it adapts to changing environments and takes advantage of new opportunities. Of course, for those of us who are palaeontologists, this 'dandruff' is precious beyond words. These parts of the Bioblob that it discarded at various stages back through time are the only things that can enable us to reconstruct the Bioblob's earlier forms to understand how it has been changing over the billions of years it has been alive. Of course, fossils can only provide an extremely limited understanding about how the whole super-organism that produced these bits has been changing over the last few billion years. Most of the organism over time hasn't shed examinable bits, or if it has, these haven't yet been discovered. Further, those that have been shed are rarely preserved. Further still, those that have been preserved are in the main hidden too far below the surface for us to find them. ●

Moments in time when the tree of life was massively pruned

MIKE ARCHER

OVER THE LAST 500 MILLION YEARS, and no doubt many times before that, there have been at least five episodes of severe 'stunting' of growth of the branches of the Tree of Life (aka the Bioblob). We call these events 'mass extinctions'. At these times, between ~75% (e.g. the end of the Cretaceous when the Chicxulub meteorite struck the Yucatan Peninsula) and 96% (e.g. the end of the Permian) of the branches of the Bioblob stopped expanding in time and space – they become 'extinct'. The most commonly recognised five 'mass extinction' events occurred at the end of the Ordovician, end of the Devonian, end of the Permian, end of the Triassic and end of the Cretaceous.

While the causes of these five mass stunting events are often controversial (ref. 321), common explanations include climate change, volcanic events, meteorites, pollution and environmental fluctuations in elements such as selenium and nickel.

The late Pleistocene saw a similar episode of savage pruning of the Bioblob, but it was a less severe pruning than the five earlier events. Hence, it is not generally regarded as a 'mass extinction' event. There has been controversy about whether a meteor impacted Earth during or preceding these Pleistocene extinctions, but it is now generally conceded that one did in fact hit somewhere in the Northern Hemisphere, possibly correlated with an abrupt climate change interval called the Younger Dryas Event that spanned 12 900–11 600 years ago. The causes of extinction on all land masses of some of the Pleistocene megafauna (i.e. animals larger than ~44 kg in weight) are still uncertain, despite being closest to us in time. It was often assumed that early humans must have been responsible for the extinctions, but it is becoming increasingly clear that late Pleistocene climate changes are at least an equally if not more credible common denominator of the losses around the world on all continents including Australia.

There are many thousands of bones of extinct late Pleistocene megafaunal species in Australia, and not one provides credible evidence that even a single individual megafaunal animal was killed by a human. In fact, as was the case in the Northern Hemisphere, much of Australia's extinct megafaunal species were gone before humans arrived on the continent (ref. 464). Discovery of what was reported to be a 34 000 years old skeleton of *Zygomaturus trilobus* at Lake Willandra, New South Wales and similar-aged megafaunal species from Cuddie Springs in New South Wales, indicate that humans and some of the extinct megafauna coexisted for at least 17 000 years. Although there is no hard evidence that humans had a significant role in the loss of any of the extinct megafaunal species, no doubt the issue will remain controversial as new evidence continues to challenge entrenched assumptions on both sides of the debate. And that evidence is coming from large investigations of sites across Australia.

Unfortunately, we humans have triggered what is now commonly regarded to be the onset of the world's Sixth Mass Extinction, with millions of animals and plants now in danger of vanishing forever. Although we could slow this unfolding process by stopping widespread habitat and climatic destruction, we don't seem to be doing enough to avoid even more horrific consequences than occurred following the end of Cretaceous mass extinction. It seems the world is in for another rough ride and possibly even our own extinction. ●

Australian palaeontology: prehistoric beginnings and growing fast

Palaeontology has been practised in Australia for thousands of years, long before the arrival of Europeans. It is clear that the first palaeontologists in Australia were Indigenous Australians. Both invertebrate and vertebrate fossils have been valued by various groups of Aborigines. Some prized fossils were transported via trade routes that spanned the continent. Among people who knew the modern biota well, it is no surprise that the remains of extinct animals inspired their keen interest in fossils. But without a written record, we know far too little about the extent of these earliest investigations.

Following European settlement in the 1700s, records of research focused on the prehistoric creatures of Australia, New Guinea and New Zealand began to accumulate and influence global understanding about evolution in the Southern Hemisphere. Even Charles Darwin was profoundly impacted by the realisation that fossil deposits in Wellington Caves, New South Wales, contained extinct wombats, kangaroos and other characteristically Australian types of animals, rather than elephants, rhinoceroses and other animal types found in the fossil deposits of Europe. He considered this was clear evidence that living animals on isolated lands had descended from local ancestors – a notable departure from the creationist belief in a universal global biota that was created de nova in the Garden of Eden before being drowned and fossilised in the sediments of a global flood. The Wellington Cave fossils were a major reason why Darwin conceived his Law of Succession of Types in 1837, leading to his 1859 publication of *The Origin of Species*.

Some of Australia's colonial palaeontologists had to struggle to demonstrate their capacity to compete for attention on the global stage. For example, Gerard Krefft in the Australian Museum, despite being one of its first and most capable curators, was instructed by his anglophilic Board of Trustees to turn over the fossils he found, along with his research notes about them, to Sir Richard Owen in the Natural History Museum in London, so that Owen had the privilege of publishing the first papers about these extinct Australian animals. Although Krefft complied at first, understandably this requirement eventually frustrated him. It triggered a long and fascinating war of disagreement between Krefft and Owen about, for example, the presumed diet of the Marsupial Lion, *Thylacoleo carnifex* – Owen argued it was a carnivore and Krefft that it was a herbivore. Krefft's refusal to bow to English authority put his job in jeopardy, but it also made him one of Australia's first scientific freedom fighters. His eventual refusal to obey the Board's instructions was

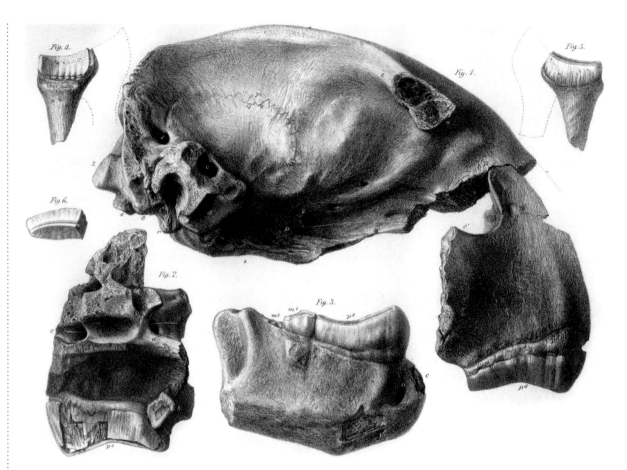

Fig. 3. One of the beautiful illustrations of extinct Australian mammals from Owen's 1877 magnum opus on the Pleistocene fossil mammals of Australia. In this case, a partial skull and teeth of the marsupial lion, *Thylacoleo carnifex*.

an important step on the path that enabled Australian palaeontology to develop in its own right – even if, as it eventually turned out, Krefft was wrong and Owen right about the diet of Thylacoleo!

In addition to the centuries of palaeontological research by many local palaeontologists, international palaeontologists have also made major contributions to our understanding about the fossil record of this continent. They include palaeontologists from the University of California, Field Museum in Chicago, University of Texas, Natural History Museum in London and many others. Of particular note in relation to vertebrates, Owen's magnificent compilation of Australia's Pleistocene animals, enabled in part by the fossils begrudgingly sent to him by Krefft, stands head and shoulders above all the others. He produced the beautifully illustrated *Researches on the Fossil Remains of the Extinct Mammals of Australia; with a Notice of the Extinct Marsupials of England* in 1877 (Fig. 3). Similarly, he produced *Memoirs of the Extinct Wingless Birds of New Zealand* (1879), a compilation of 40 years of research describing the fossil birds from New Zealand.

Palaeoartistry: visionary bridge between palaeontology and the public

While palaeontologists have relatively little difficulty in comprehending the significance of most of the species they discover, sometimes based on little more than a leg bone or a tooth row, understandably the public needs more help to bridge that visualisation gap. This is why palaeoartists are so important. Translating new discoveries into whole-body reconstructions (Fig. 4) requires multifaceted skills. Rendering living animals as paintings is relatively unchallenging, as the whole animal is available to use as a model. However, to reconstruct extinct vertebrates, the artist must have more than well developed artistic skills; they must have deductive skills to work out how novel morphological features of the bones, muscle attachments and teeth translate into differences in overall shapes and behaviours. It helps if the artist also understands the nature of the ecosystems within which the animal lived,

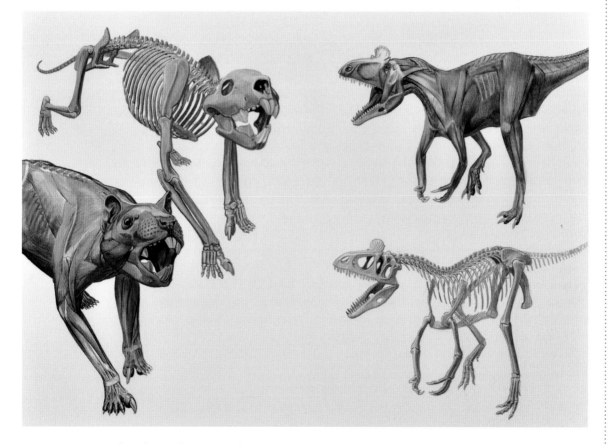

Fig. 4. Before a palaeoartist can produce a plausible reconstruction of an extinct vertebrate such as a Marsupial Lion *Thylacoleo carnifex* (left) or an Antarctic dinosaur *Cryolophosaurus ellioti* (right), they need to work with palaeontologists to determine the approximate size and orientation of muscle masses attaching to the different elements of the skeleton. Often this understanding can be deduced by the size and shape of 'muscle scars', roughened areas where the muscles originated or inserted onto bone surfaces.

because these provide further constraints in relation to such things as plausible body markings and overall colour. For example, modern vertebrates that live in water tend to have darker upper and lighter lower colours while vertebrates that live in forests often have patchy or striped body markings that help to hide them from potential prey as well as from predators. While some palaeoartists embrace the opportunity to produce animals such as dinosaurs in glorious technicolour, arguing that some of their descendants such as birds have such colouring, others take a more conservative approach. This often leads to radically different renderings of the same animal. For most of the animals reconstructed by Peter Schouten in this book, there have been no previous full-bodied reconstructions so there will not be too many conflicts of interpretation – yet!

There will continually be new interpretations, based on new materials about the overall body shape and lifestyles of the animals being reconstructed. A classic example has been the revolution in our understanding about almost everything to do with one of the most common fossil marsupials found in the middle Miocene of Riversleigh: *Nimbadon lavarackorum*. When this was first discovered, all we had were a couple of tooth rows. From these it was deduced that this mammal was a new kind of herbivorous diprotodontid similar to others previously found, such as species of *Ngapakaldia* or *Zygomaturus*. It seemed probable that it was a quadrupedal leaf-eater that browsed on low vegetation growing in the understorey of the Miocene rainforests. But in the years that followed, we began to find much more of these animals, eventually including entire articulated skeletons. Karen Black began an intensive study of these skeletons and, much to everyone's surprise, concluded that just about everything we had presumed about them was wrong. The limb bones and claws indicated that rather than roaming the floor of the forest they were in fact more like sloths or sun bears in their behaviour, climbing trees and entirely capable of hanging upside down (p. 140). Further, current study of microwear on their teeth has indicated that, contrary to our earlier naïve presumptions, they weren't eating leaves; what they were eating while hanging upside down in the crowns of the rainforest trees is still under investigation. All this has added up to recognition that these were 'drop bears' in a manner of speaking – it is quite clear that now and then one of these sun bear-like marsupials would lose its grip in the trees and fall into the cave below, where we have now found the fossilised remains of more than 26 individuals from tiny juveniles to old adults.

Peter Schouten has been studying, interpreting and rendering Australia's living and fossil vertebrates for many years with consummate skills that get more stunning every year. When particularly fascinating new discoveries have been made in the Australasian fossil record (e.g. Fig. 5), Peter has produced an illustration that brings to life for the general public what it is that has so excited the palaeontologists about the new animal.

For similar reasons, many scientific journals that publish descriptions of exciting palaeontological discoveries have used Peter's art for the cover of those issues (e.g. Fig. 6).

Not uncommonly, Peter's artwork has been used to illustrate new species named in honour of individuals who have made major contributions to Australian vertebrate

Fig. 5. Peter Schouten's illustration to accompany a scientific description of *Malleodectes mirabilis*, a member of a new family of marsupials, the Malleodectidae, found in the Miocene deposits of Riversleigh that may have been specialised snail-eaters. These cat-sized carnivorous marsupials had huge, ball-peen hammer-like premolars that could only have been used for smashing hard food items such as snail shells.

Fig. 6. Reconstruction by Peter Schouten of *Obdurodon tharalkooschild* used as the cover art for the November 2013 issue of the *Journal of Vertebrate Paleontology*. Included in the cover art is an image of the first lower molar of this giant toothed Platypus and one of the fossil turtles from Riversleigh. This enormous Miocene toothed Platypus may have been more than 1 m in length and thus at least twice as long as the living Platypus.

Fig. 7. Following a long tradition of naming new species after individuals who have provided significant support for palaeontological research, this very small rainforest Marsupial Lion - illustrated by Peter Schouten - was named after Sir David Attenborough, *Microleo attenboroughi*. This cat-sized carnivore hunted in the early Miocene rainforests of Riversleigh.

palaeontology. As an example, in recognition of Sir David Attenborough's role in increasing global awareness of the fossils from the World Heritage Riversleigh area, Anna Gillespie and colleagues named a highly specialised marsupial lion after him. It was illustrated by Peter Schouten for the scientific publication (Fig. 7).

Peter's skills have long since been embraced by the rest of the world. He has reconstructed extinct and living creatures from all continents, although his primary focus has been the natural history of Australia and this region of the world. We've had the good fortune of working with Peter over many years, and the present book is one of the most comprehensive to date about this region's prehistory.

The future for discovering more about prehistory in Australasia

At a wild guess, the creatures we have focused on in this book represent less than 1% of what palaeontologists have discovered about this region's extraordinary past. At a wilder guess, all that has been discovered to date about its prehistoric menagerie probably represents less than 1/1 000 000th of 1% of the creatures that have called this place home. The challenge for future generations trying to understand more about this region's history will be to grab their hats, leave the bitumen-covered cities and start systematically searching for new fossil deposits that are out there awaiting the next generation of prehistory explorers.

Every time there's a flood, the banks and gravel bars of rivers should be examined. Whenever dams are dug in remote areas, no matter how small the hole, what's dug up should be carefully examined. When bushwalking, all non-volcanic rocks should be studied for fossil bones or tracks. Given that most fossil deposits in Australia have been found by sharp-eyed members of the public, finding new important fossil deposits is not a job that belongs only to professional palaeontologists!

The key for the beginner, whether looking for fossils or even searching for new insects, is to look for odd things in unusual places – a strange shape, curious texture or oddly coloured 'something'. In a sandstone face, for example, fossils such as bones, cones or shells are more likely to be found in ancient channel deposits revealed by discrete layers of larger pebbles. Keeping your face ~30 cm from the rock face, each object in the channel deposit should be carefully and thoughtfully examined. If an object of interest can be removed, it's important to photograph it with something to indicate the size of the object, mark the place on your map and, if possible, record GPS data for the discovery site so you know for certain the location of this exact spot. Then take the object or its photo to a palaeontologist in the nearest museum or university. If you find a skeleton in a cave, take a close-up photo on your phone (include a scale) and email it to a palaeontologist in one of Australia's museums or universities, with all the information about the location that you can relay.

Given that landowners in rural and regional areas, bushwalkers and cave explorers almost certainly outnumber palaeontologists by at least 1000 to 1, it is easy to understand why most important discoveries about our prehistoric past have been made by observant, curious and interested members of the public rather than by the palaeontologists themselves.

It's time to wander through the pages of Australasia's prehistoric record, with Peter Schouten's skills to bring these creatures back to life.

Pilbara, Western Australia

ARCHAEAN

HOW DID LIFE FIRST BEGIN ON EARTH? SOME have argued, given the fact that stony meteorites all over the world have been found to contain amino acids and other organic molecules, that Earth's life may have had extraterrestrial origins. While it is probable that life abounds elsewhere in the universe, hard evidence for ancient, very simple life forms here on Earth is also abundant. Some of the most interesting comes from sedimentary rocks about 3.5 billion years old exposed in the Pilbara district of Western Australia, near a town called Marble Bar. Rocks here contain minerals that suggest they formed in hot springs where synthesis of Earth's first life may have begun. Here, there are also fossils called stromatolites, structures built by unicellular organisms called cyanobacteria. Because simple cells of this kind still occur in extreme environments in the modern world, and in some cases still build stromatolites such as those found today in the hypersaline waters of Shark Bay, Western Australia, we know a lot about them. They can harness solar energy via photosynthesis to convert inorganic atmospheric gases into organic compounds, the chemical building blocks they need to thrive.

In this reconstruction of a scene 3.5 billion years ago, the sky appears orange rather than blue because at this time in the early history of the atmosphere it contained mostly greenhouse gases and almost no free oxygen. The shallow ocean waters were green because they were chemically different from oceans of today; altogether it was an eerie planet, but one where exciting things were happening. At the edges of the shallow ocean in a volcanically active world, hot springs were bubbling up with mineral nutrients that could have enabled

AGE

Archaean Eon, about 3.45-3.43 billion years ago.

LOCALITY

Pilbara, Western Australia (Dresser Formation). Possibly even older stromatolites have been reported from rocks in Greenland that are 3.7 billion years old.

ENVIRONMENT

Shallow sea adjacent to an area of land where hot springs were disgorging water and mineral nutrients from rocks below the surface. The springs would have produced pools of varying temperature where life may have begun. In the shallow ocean waters, stromatolites developed as mounds built by cyanobacteria that faced the sun. Volcanoes were active at this time, providing a constant source of nutrients.

REFERENCES

6, 100, 280

the first life to develop. In the shallow marine waters nearby, stromatolites are busily converting CO_2 into more complex organic compounds and giving back O_2 as a waste product. They were the source of the gradually accumulating mass of oxygen that eventually enabled animals, such as the vendozoans found at Ediacara in South Australia (p. 20) to evolve, thrive and diversify. Because of this fossil record in Western Australia, NASA's efforts to find evidence of former life on Mars are increasingly focused on investigating what appear to be ancient Martian spring deposits. ●

Ediacaran Hills, South Australia

EDIACARAN

APPROXIMATELY 560 MILLION YEARS AGO, THE southern Flinders Ranges of South Australia near Brachina Gorge in the Ediacara Hills was nothing like the dusty desert landscape it is today. At that time, a shallow warm sea covered the area and in it thrived many of Earth's earliest animals. Some of these were trapped in fine silt that eventually buried them, enabling the process of fossilisation. Discovered by geologist Reg Sprigg in 1946 and now famous worldwide, the Ediacaran fossils are unique in their high diversity and excellent preservation. They represent a rare window into the appearance and lifestyles of complex multicellular animals on Earth. The time interval when these creatures lived has been formally named the Ediacaran period. The scene shown here contains feather-like sea pens (pennatulaceans) like *Charniodiscus* to the right. Like corals, these filter the water for organic matter. The little polyp-like creatures that live on each 'leaf' are akin to living coral organisms. Some of these sea pens were up to 50 cm in height and each was attached by a thick basal plate to the substrate. The thimble-shaped animal with long spikes radiating out was perhaps the world's first creature to have a supporting skeleton. Named *Coronacollina*, it was only about 1.5 cm high but its spicules extended up to 37 cm long. It likely fed like a sponge, sifting organic material form the water.

The colonial organisms in the background which are dividing into strands are *Funisia*, perhaps the earliest evidence of sexual reproduction in the fossil record. The fact that they budded at the same time rather than randomly suggests a mass mating event somewhat like the spawning of corals under a full moon. The giant flat worm on the bottom right is one of the largest known organisms of this age, it is the *Dickinsonia rex*, just over 1 m in length. It slowly moved over the sea floor either absorbing nutrients through its skin or grazing, depending on whether or not it had a mouth. Smaller *Dickinsonia costata* is shown in the centre of the image. Discovery of cholesterol in similar-aged species of *Dickinsonia* in Russia confirms that these are in fact archaic animals because only animals have cholesterol. Other creatures depicted here include the enigmatic *Tribrachidium*, centre right, and *Kimberella*, the small elongate animal at bottom centre. ●

AGE

Late Ediacaran Period, about 570-540 million years ago.

LOCALITY

Ediacara Hills, Flinders Ranges, South Australia.

ENVIRONMENT

Shallow warm sea floor is the most accepted interpretation, ranging from the marine zone of fair weather wave oscillation through to distal storm wave base. A recent paper (ref. 323) suggested that some of the Ediacarans may have inhabited a cool dry soil environment, but this view has been debunked in detail (ref. 380).

REFERENCES

77, 103, 238, 323, 364, 380

Emu Bay, South Australia

LIKE ITS SLIGHTLY YOUNGER COUNTERPART THE Burgess Shale in British Columbia, and the slightly older Chengjiang Fauna in Yunnan, China, the formation called the Emu Bay Shale is a world-class deposit renowned for its high quality of fossil preservation. Here, not only the hard-shelled arthropods like trilobites (centre right and left) but also many kinds of soft-bodied worms and creatures of enigmatic affinity were exquisitely preserved between layers of black shale.

The top predator in this ancient ecosystem was the arthropod *Anomalocaris* (centre top) represented by two species in the fauna, the endemic *A. briggsi* and *A.* cf. *canadensis*, also known from the famous Burgess Shale site in Canada. *A. briggsi* grew to about 50 cm in length. It is seen here with a small arthropod *Kangacaris*, a trilobite-like form, struggling in the big predator's appendages and about to be eaten. Although *A. canadensis* is interpreted as a feeder on hard-shelled animals like trilobites, its Australian cousin was less well equipped for hard food, and probably fed mainly on small arthropods and soft-bodied creatures. *Anomalocaris* had large pear-shaped eyes on stalks with about 16 000 lenses on each eye, giving it the most acute vision of any creature for this age, and equivalent to the visual acuity of a modern dragonfly. Its long feeding appendages could grasp and snatch unwary prey and lift it to its deadly circular saw-like mouth. The large trilobite in the bottom left corner is *Redlichia takooensis*, one of the commonest fossils found at Emu Bay, leaving a *Cruziania* feeding trail in its wake.

Bivalved arthropods, the earliest known prawn-like creatures, swim to the top left of the scene. *Isoxys* can be seen in the top left and *Tuzoia* a little below it. A nudibranch mollusc is seen swimming just below the centre of the scene. Although nudibranchs haven't been found at Emu Bay they are likely to exist at this time, as the first molluscs were radiating into the major groups found today. ●

AGE
Early Cambrian, about 514 million years ago.

LOCALITY
Emu Bay Shales, quarry on the north side of Kangaroo Island, South Australia.

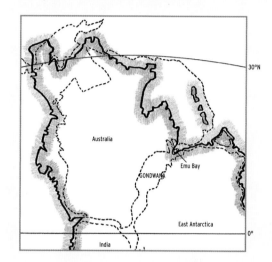

ENVIRONMENT
The south-western corner of what is today Kangaroo Island was underneath a warm tropical seaway close to the ancient equator. Turbidity currents caused by tectonic activity would periodically wash animals down the shelf slopes into a deeper mini-basin on the continent's inner shelf, preserving very fine details of each fossil.

REFERENCES
105, 137, 211, 287, 288, 289, 290

Stairway Sandstone, Northern Territory

ORDOVICIAN

ABOUT 465 MILLION YEARS AGO A SHALLOW warm inland waterway called the Larapinta Sea cut Australia in half and soaked central Australia, then part of the larger Gondwana landmass. In these tropical waters lived some of the earliest known backboned animals (vertebrates), but they did not look like any of today's fishes. *Arandaspis* (the five individuals at right), named after the Aranda Indigenous peoples of the region and Greek 'aspis', meaning 'shield', is known from relatively complete impressions of its bony head shield and elongated trunk scales. It bore two small headlamp-like eyes at the front of its head and a small open mouth, lacking jaws, probably for sifting organic particles out of the sandy sea floor. Growing to about 25 cm in length, it was invested with finely sculptured bone formed in the skin (dermal bone) over its entire outer surface, but apparently lacked any ossifications inside it. It thus highlights the peculiar nature of our deep distant evolution – the first vertebrates were 'inside out' animals compared to later more advanced creatures, like us, where all the bone is deep inside the animal.

Also closely related to our ancestry are the

tiny worm-like conodonts, extinct chordates with phosphatic jaw-like structures inside the head. Some scientists place conodonts as more advanced than fishes like *Arandaspis* because the presence of phosphatised tissues similar to dentine is a feature that would later characterise all jawed animals. Others place conodonts a step lower on the evolutionary tree than jawless fishes because we simply don't know enough about them yet, as few are ever preserved whole. A number of them are seen here homing in on the wake of the trilobites' feeding trail. Not shown here is *Porophoraspis*, a close relative of *Arandaspis* that had a different kind of surface ornament on its bones. Scales of a tantalising possible early jawed shark-like fish, named *Tantalepis*, have been found in the same deposits. They are exactly like the simple placoid scales seen in many living sharks.

Giant orthocone nautiloids, squid-like creatures living inside long straight chambered shells, were clearly the top predator in this shallow marine ecosystem, some growing as large as 2 m in length. They would likely be feeding on large trilobites and other sedentary molluscs they could catch easily. Huge trilobites, some up to 75 cm in length, are represented in the fauna by their feeding trails, called *Cruziania*, which characterise the outcrops of the Stairway Sandstone. Trilobites feed by pushing organic-rich sediment into their backwards-facing mouths, thus leaving the distinctive trail from the scraping motion of their feeding appendages. Two examples of the little bivalve *Aloconcha* can be seen lying in the wake of the giant trilobites' feeding trail. ●

AGE

Middle Ordovician (Darriwilian), about 465 million years ago.

LOCALITY

Stairway Sandstone, Mt Watt, Mt Charlotte, central Australia.

ENVIRONMENT

Warm equatorial tropical inland sea. Australia then formed the northern edge of Gondwana.

REFERENCES

100, 339, 346, 383

Baragwanathia flora, Victoria

SILURIAN/DEVONIAN

SOME OF THE WORLD'S OLDEST WELL-PRESERVED land plants are represented by fossils found in central Victoria near the town of Yea. They are dated at most late Silurian age, around 420 million years old. This makes *Baragwanathia* one of the most advanced land plants for its age. Fossils of *Baragwanathia* persist until early Devonian times about 395 million years ago. These include the clubmoss (lycophyte) *Baragwanathia longifolia* shown here as the central slender branching plant. It grew up to 2.5 m in length and each stem was around 5 cm in diameter. It was named in honour of the former director of the Victorian Geological Survey, William Baragwanath. It was a vascular plant with supporting tissues in the stem, each of which bore a loose spiral of short spiny leaves. It had adventitious roots that crept down from the leafy stems to anchor it to the ground. It is thought to have lived alongside rivers and marshes, where occasional floods swept the plant out to sea.

Baragwanathia was discovered and named in the mid 1930s by Isabel Cookson. An early plant species, *Cooksonia*, was later named after her. It is shown here as the clump of erect branching stems emerging from the water, each bearing terminal sporangia (to the right of the central *Baragwanathia* plant). *Cooksonia* was perhaps a little more primitive than *Baragwanathia* because it has a simpler type of vascular tissue supporting the stem. It grew only a few centimetres tall and lacked leaves. It belongs in the Embryophyte group which contain mosses, ferns, liverworts and their kin. Some *Cooksonia* stems contain holes that aided gas exchange (like leaf 'stomata') but the plant was not capable of photosynthesis. It is now known from about half a dozen different species around the world. To the left of the central *Baragwanathia* plant are some green weakly branching stems emerging upright from the water. These are zosterophyll plants of unknown identification. These simple branching plants have stems covered with very short spines and grew by unrolling from the tips. Other plants depicted here are known from their fossil spores in rocks of similar age in southern Australia. They allow us to reconstruct clumps of early ferns (right, centre), mosses (centre left, top) and liverworts (behind the central *Baragwanathia*). These are all representative of living forms found in Australia and elsewhere today.

What animals might have lived among the early plants of Victoria? Although no body fossils of animals have been found in these beds, fossils of this age associated with similar plants are known from other sites in the Northern Hemisphere. They include creatures such as tiny spider-like trigonotarbids, a few millimetres long, and early insect-like springtails that possibly fed in the leaf litter and debris of the plant undergrowth. One specimen of a fossil fish has been found in the *Baragwanathia* beds near Yea. Named *Yealepis*, meaning scale from Yea, it was a small shark-like fish covered in tiny scales. Because its head is unknown, its affinities to any modern fish remain a mystery. ●

AGE

Latest Silurian-early Devonian. First sites asserted as late Silurian, later found to be early Devonian. Later, definite Silurian *Baragwanathia* was confirmed.

LOCALITY

Wilson Creek Shale, Yea, central Victoria.

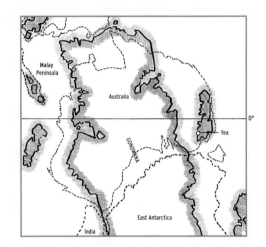

ENVIRONMENT

The fossil plants were washed out to sea and deposited in a deep marine environment. At this time the landmass was not far away, so the plants are thought to have lived in waterways close to the river mouth from where they could be swept out to sea by flood events.

REFERENCES

68, 85, 102, 138, 206

Evolution of early vertebrates

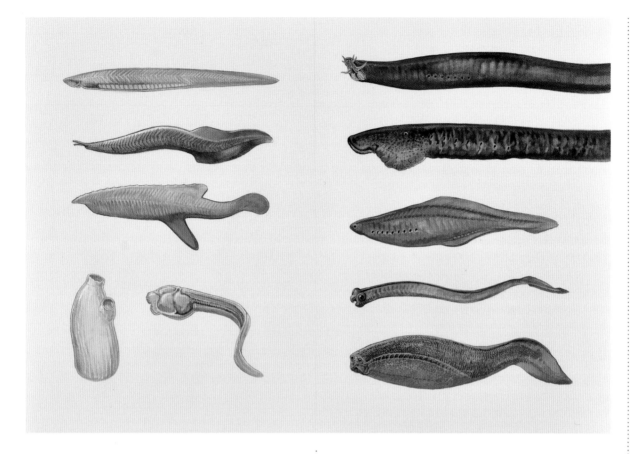

THE EVOLUTION OF VERTEBRATES HAD HUMBLE beginnings with simple worm-like creatures whose main characteristics are the presence of a cartilaginous rod along the back, called a notochord. Top left we see an extant cephalochordate *Branchiostoma* (also known as *Amphioxus*). Second from the top is *Pikaia*, which lived 508 million years ago. Its fossils have been found in the Burgess Shale of British Columbia. It had W-shaped muscle bands, a dorsal notochord or stiffening rod, possible gill slits and a head with two well-developed eyes. These features are among those that make it a member of the Phylum Chordata and a likely contender as a vertebrate ancestor. Below *Pikaia* we see *Palaeobranchiostoma*, a fossil tunicate from the Permian of South Africa. Tunicates or urochordates include the modern sea-squirts (below left, sessile

adult sea squirt; right, mobile larval form). The muscular body of the larval form has been the basis for reconstructing *Paleobranchiostoma* as a chordate with fins that had the ability to swim. Its dorsal fin had several strong barbs along its leading edge. Sea squirts (bottom left) have a free-swimming larval phase that eventually settles down on the sea floor where it becomes a sessile adult that filter-feeds from the water. The presence of a well-developed pharynx with gill slits and a muscular tail places sea squirts close to the ancestry of all vertebrates, and in recent years molecular analysis of their DNA has confirmed that this is so.

The hagfish *Myxine* (top right) is perhaps the most primitive living fish, although strictly speaking most biologists do not regard it to be a true fish anymore; after all, it lacks jaws, teeth, fins and quite a few other anatomical features seen in all modern fishes. Perhaps *Myxine* would be better regarded as an evolutionary step between urochordates and 'real' fishes.

The lampreys (right, second from top) are parasitic jawless fishes, known from many species. They have an ammocoete juvenile phase that filter-feeds in the substrate. The adult has a migratory phase in the open sea, and many species return to rivers to spawn. They all have a well-developed oral hood with keratinous horny teeth to rasp their prey and suck blood. Their gills are internally lined in pouches, unlike jawed fishes which have their gills supported on lateral bars. Fossil jawless fishes like *Myllokungmingia* (centre right) show the presence of a well-developed head, well-developed gill rows, and muscular tail. While their 530 million years old fossil remains from Chengjiang, China, are difficult to interpret, it is likely that they were the first true fishes, although their skeletons still lack ossified bone. *Mettspriggina*, recently described from the Burgess Shale, is a close contender for the first true fish because it has well-developed gill-arches supported on lateral gill bars like jawed fishes.

The conodont animal (second from bottom, right) has phosphatic bony tissues forming its feeding apparatus which looks jaw-like and contains a dentine-like layer. This is a true vertebrate characteristic, a tissue derived from neural crest cells that occur in all living animals. Complete conodont remains of *Clydagnathus*, dated around 330 million years old from Scotland, reveal a worm-like body with large eyes and a simple tail. Finally (bottom right) we see an early jawless fish, *Arandaspis*, from the 465 million years old Stairway Sandstone of central Australia, with well-developed bone covering its entire external surface. The next biggest phase in vertebrate evolution was to acquire jaws and teeth. Early shark-like scales found with *Arandaspis* suggest this could have taken place by the end of the Ordovician Period. ●

REFERENCES
97, 182, 264, 355, 359

Burrinjuck, New South Wales

DEVONIAN

THE DARK BLUE LIMESTONES AND LIMEY SHALES outcropping in the Taemas-Wee Jasper region of New South Wales around Lake Burrinjuck were laid down in a warm shallow sea about 410–400 million years ago. Here, patches of algal reefs with interspersed colonial corals have produced the most diverse and best-preserved fauna of early fishes of this age, anywhere on the planet. The specimens are acid-prepared from the rock to reveal stunning 3-D skulls and bones that have been recently studied using micro-CT imagery to reveal the internal anatomy down to the cellular level. At least 70 species of fishes are now known. These include armoured placoderm fishes, the first vertebrates with jaws and teeth, as well as many kinds of early bony fishes. Isolated jawbones, teeth and scales of spiny acanthodians ('stem chondrichthyans') and jawless thelodonts are less common in the fauna.

The scene here shows the largest fish from this habitat, the placoderm *Cavanosteus*. It was probably a slow-swimming filter-feeder that swam like a whale shark through the warm waters of the Gondwana seaways, feeding on plankton. Below left, resting on the coral, is an acanthothoracid placoderm *Murrindalaspis* with its high median dorsal spine. Just behind it, lurking under the coral is a large lungfish *Dipnorhynchus*, which probably fed primarily on hard-shelled clams and snails. To its right are two more placoderms, these with flared spinals in front of the fins. The one on the right is the petalichthyid placoderm *Widjeaspis*. In the background an eel-like bony fish *Onychodus yassensis* darts out from the reef to catch an unwary passing placoderm.

Another fish in this fauna was *Dhanguura johnstoni* (centre right). The generic name means 'catfish' in the local Wiradjuri Indigenous language, while the species name honours its finder Dr Paul Johnston. *Dhanguura* may have reached lengths of about 2–3 m. We do not have its jaw bones so have no idea what it might have eaten. In the bottom right of the image we see a school of three ptyctodontid placoderms cruising into the scene. These are durophagous (eating 'hard' foods) fish that have powerful tooth plates for eating hard-shelled prey on the reef. The body restorations shown here are based on better-known forms from the late Devonian Gogo Formation because only their tooth plates have been described from the Taemas site.

In the far distant mid-surface waters we see silhouettes of other placoderms in the centre, and some early possible ray-finned fishes (actinopterygians) to the right. We have isolated scales and bones suggesting these were present, including a skull of one form, *Ligulalepis*, a primitive kind of osteichthyan. ●

AGE

Early Devonian, from Pragian–Emsian, about 410–400 million years ago.

LOCALITY

Around Burrinjuck Dam, Taemas-Wee Jasper region, New South Wales.

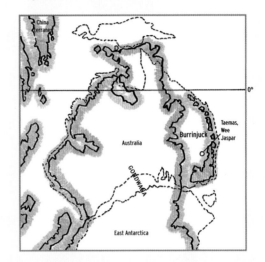

ENVIRONMENT

Abundant marine invertebrates like shells, corals, trilobites, brachiopods and crinoids in the fish-bearing strata confirm this was largely a shallow tropical reef environment.

REFERENCES

213, 227, 386, 467, 468, 469

Georgina Basin, Queensland

DEVONIAN

THE GEOGRAPHIC REGION FROM THE CENTRAL western Queensland border to the Canning Basin of Western Australia and down to Cobar in western New South Wales was once inhabited by the *Wuttagoonaspis* fauna which contained strange, armoured fishes and rare bony fishes. *Wuttagoonaspis* was a peculiar placoderm (named after Wuttagoona Station in New South Wales) that grew to 1 m in length. It is shown here as the largest fish (to the left of centre) with two kinds of peculiar jawless armoured fishes glancing at it from either side. *Wuttagoonaspis* is an enigma in

the world of placoderms. It shows features of skull roof and body armour that are seen in the most diverse group of placoderms, called arthrodires, but if it belongs in this group at all it is very basal. Its remains are highly distinctive fossils characterised by linear patterns of ornamentation on the bones. Swimming away at the left of the scene is *Neeyambaspis*. The fish on the right with the long snout is the very distinctive *Pituriaspis*. Indeed, *Pituriaspis* and its cousin *Neeyambaspis* belong in a group called the Pituriaspida. When Gavin Young erected it in 1992, it was the first new class of vertebrates discovered in over 70 years. We know very little about them because they are only known from a few sandstone casts in the rocks.

In the background is a flattened placoderm typical of the early phyllolepid types, possibly one of the rare arthrodires in the fauna. Phyllolepids emerged as a common group later in the middle Devonian. This occurrence in the Georgina Basin sediments could well be the oldest known example of the group. The *Wuttagoonaspis* fauna is also famous for containing the oldest known members of the rhizodont fish group, which went on to become the largest bony fishes of the Palaeozoic (later forms reached 6 m in length). The Cravens Peak rhizodontid was small, about 50 cm long, and is known only from its shoulder girdle and a few odd bones. An isolated limestone layer within the succession bearing these fish fossils has yielded teeth of sharks (*Maiseyodus*) as well as scales of acanthodians (basal shark-like fishes) and tiny jawless fishes called thelodonts. A single jaw also shows that the eel-like onychodontids were present (*Luckius*).

The most fascinating thing about the *Wuttagoonaspis* fauna of Australia, known from an area of nearly 2 million km^2, is that it was this country's first widespread endemic vertebrate fauna. Around 30 species are now known from this fauna, the vast majority of which are placoderms. Many new specimens are being studied, so this number is bound to expand in the future. ●

AGE

Early-middle Devonian Boundary, about 395-390 million years ago.

LOCALITY

Craven Peak Beds, Georgina Basin, western Queensland; Dulcie Range, Northern Territory; and Cobar, New South Wales (Mulga Downs Group). Subsurface deposits containing the fauna are also known from the Canning Basin in Western Australia and Officer Basin in South Australia.

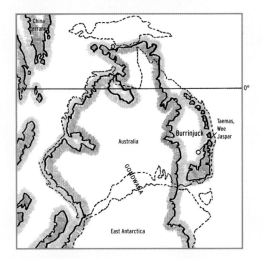

ENVIRONMENT

The lack of marine invertebrates and microfossils in the fish-bearing strata suggest this was largely a fluviatile environment, but the presence of thelodonts, common in marine deposits in the early Devonian, suggests the sediments could have been deposited close to the mouth of the river system.

REFERENCES
335, 336, 398, 466, 470

Gogo, Scene 1, Western Australia
DEVONIAN

THE GOGO FORMATION FORMED IN THE INTER- reef basins between active reef fronts and atolls in a warm tropical sea close to the equator about 382 million years ago. The reef was built up of several kinds of algae and layered sponge-like stromatoporoids with scattered corals and bryozoans. Clusters of horn-shaped rugose corals and large colonies of tabulate forms are shown here as well. Many kinds of fishes lived here, and this scene features some of the more bizarre bony fishes (Osteichthyes) as well as some of the placoderms. The perfect 3-D preservation of the Gogo fish fossils, including muscles and soft tissues in some specimens, makes it the world's best preserved and most diverse assemblage of fishes of this age. Studies on the geology of the site indicate the fishes' bodies sank into a quiet basin whose muddy seafloor lacked oxygen. Middle water layers rich in hydrogen sulphide helped preserve the animals' delicate internal structures when the fossilisation process began. Today their bodies, entombed in limestone concretions, can be found weathering out in outcrops of the Gogo Formation east of Fitzroy Crossing.

The long-shielded placoderm *Holonema westolli* (top left) was a bottom-feeder, probably browsing on algal oncolites with it scoop-shaped dental plates. It grew to about 1 m long and is known from several species around the world, but *H. westolli* from Gogo is the best preserved. Some specimens show pebbles in the gut, suggesting it either ate oncolites or ingested ballast stones to keep it on the sea floor while feeding. At the centre lower part of the scene we see the long snouted placoderm *Rolfosteus*, whose crushing tooth plates suggest it ate hard-shelled creatures like the bivalved concavicarid crustaceans that inhabited the sea floor. Its streamlined body with long tubular snout must have been useful for keeping ahead of predators in this highly diverse ecosystem.

To the bottom right we see the dagger-toothed fish *Onychodus jandemarrai*, named after Jandemarra, an early Indigenous freedom fighter who lived in the region. This fish, which grew to 2 m long at Gogo but up to 4 m elsewhere, was like a killer moray eel that darted in and out of the crevices in the reef to catch unsuspecting passing prey. In this case it has seized a long snouted lungfish *Griphognathus whitei*. *Griphognathus* likely used its elongate duck-like jaws to suck up worms and other prey in the muddy sections of deeper sea floor, perhaps using its electro-sensory system at the tip of the snout. Small ray-finned fishes *Mimipiscis* swim in the top background centre as does a small arthrodire, *Compagopiscis* (top right). ●

AGE
Late Devonian (Frasnian), about 385–380 million years ago.

LOCALITY
Gogo and Mt Pierre Stations, east of Fitzroy Crossing, Western Australia.

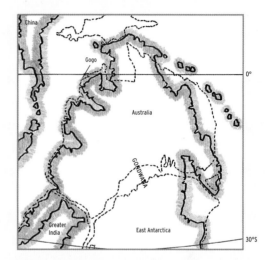

ENVIRONMENT
Warm equatorial algal-stromatoporoid reef environment, 100–200 m deep in inter-reef basins where the fishes were buried. They mostly lived on and round the reef which would have been a more active habitat, quite shallow in places.

REFERENCES
8, 73, 98, 224, 225, 249

Gogo, Scene 2, Western Australia

THIS RECONSTRUCTION SHOWS VARIOUS KINDS of placoderm fishes in the foreground with a school of ray-finned fishes *Moythomasia durgaringa*, each about 20 cm long, swimming in the top right background.

Mcnamaraspis kaprios (top left) is a 30 cm long predatory arthrodire which was proclaimed as the state fossil emblem of Western Australia in 1995, thus becoming Australia's first such emblem. Its name honours palaeontologist Ken McNamara and the species name *kaprios* (boar-like) refers to the tusks on its lower jaw. Only two specimens are known. One is in the British Museum in London and the other, a complete specimen, is on display at the Western Australian Museum.

To the bottom left we see the ptyctodontid placoderm *Materpiscus attenboroughi*, named in honour of Sir David Attenborough. It is also known as the 'mother fish' because at the time it was described its fossil remains contained the oldest known vertebrate fossil embryo, still connected to the mother by a mineralised umbilical cord. It constitutes the oldest hard evidence for live birth (viviparity) then known in any vertebrate animal and caused a stir because it meant that placoderms, once thought to be very primitive fishes, in fact had a very advanced form of mating that involved internal fertilisation. Male ptyctodontids had hook-shaped bony claspers which they inserted inside the female to transfer sperm; this was one of the earliest acts of copulation. The scene shown here hints at the imminent birth of a young pup as the tail pokes out from the mother fish's cloaca.

To the right is the strange placoderm fish *Bothriolepis*, which possessed bone-covered articulated pectoral fins. Its weak jaws suggest it fed by ingesting organic-rich muds from the sea floor. A close relative of *Bothriolepis* called *Microbrachius* was recently shown to possess bony claspers on the males which were fixed rigidly to the body plates on their underside. These fishes most likely mated side by side with arms interlocked.

In the centre background, below the *Moythomasia* shoal, is the high-spined ptyctodontid *Campbellodus*. Its body was covered in large overlapping scales. This one is clearly a male because of the claspers seen poking out behind its pelvic fins. ●

AGE

Late Devonian (Frasnian), about 385–380 million years ago.

LOCALITY

Gogo and Mt Pierre Stations, east of Fitzroy Crossing, Western Australia.

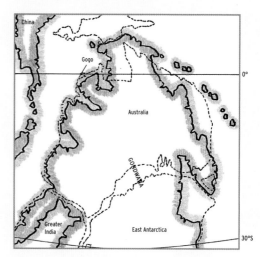

ENVIRONMENT

Warm equatorial algal-stromatoporoid reef environment, 100–200 m deep in inter-reef basins where the fishes were buried. They mostly lived on and round the reef which would have been a more active habitat, quite shallow in places.

REFERENCES

74, 218, 219, 224, 226

Canowindra, Scene 1, New South Wales

ABOUT 363 MILLION YEARS AGO IN CENTRAL
New South Wales near the small town of
Canowindra a great drought occurred. It caused
many fishes living in the large meandering river
systems to pool together and die in the last
vestiges of the water. Subsequently, layers of
water-borne sands buried them, eventually
creating a spectacular assemblage of fossils.
The bone eventually weathered away, leaving
clear impressions of their skeletons in the silty
sandstone layers. In the mid 1950s a road cutting
put through near Canowindra unearthed the
first slab covered in fish fossils and more recent
excavations by Dr Alex Ritchie, then of the
Australian Museum, and his colleagues unearthed
many new species of fishes from the site.

The scene here shows the pool just before
all the water has dried up. It mostly comprises
hundreds of examples of the antiarch placoderm
fishes *Bothriolepis yeungae* and *Remigolepis
walkeri*. *Bothriolepis* (pitted scale), which has
long segmented pectoral fins, is one of the
most common types of placoderms – about 150
species are known worldwide from both marine

and freshwater deposits. *Remigolepis* (oar scale) has a single unsegmented bony arm forming its pectoral fin. It is seen here with a slightly reddish tinge, whereas *Bothriolepis* is shown as being green. These antiarchs most likely fed by ploughing into the river bottom, ingesting organic-rich muds because their mouths had very delicate weak jaws that lacked real teeth.

The lobe-finned fishes in the centre are *Canowindra grossi*, first described from the site in 1973 and still only known from one specimen which sits in the middle of the original 1956 slab of fishes. *Canowindra grossi* is named after the township and the famous German palaeontologist Walter Gross. It belongs to the tetrapodomorph fishes, which are the lineage of fishes that is closest to early land animals (tetrapods). *Canowindra* represents an endemic family of this group only known from East Gondwanan Devonian deposits of Australia and Antarctica. These fishes had powerful jaws and strong muscular fins and would have been strong enough to gain pole position in the centre of the remaining water supply. ●

AGE

Late Devonian (Famennian), about 363 million years ago.

LOCALITY

Near the township of Canowindra, New South Wales.

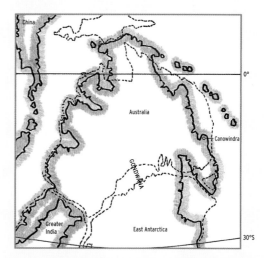

ENVIRONMENT

Fluviatile, river deposit overflow (billabong) that has dried up, leaving the fishes to die exposed. At this time Australia was further north than today, and slightly warmer with higher levels of oxygen creating lush forests near the waterways.

REFERENCES

189, 190, 215, 222, 337

Canowindra, Scene 2, New South Wales

THE ANCIENT RIVER SYSTEM NEAR CANOWINDRA was home to a diverse community of freshwater fishes that included armoured placoderms *Bothriolepis yeungae* with its bony articulated arms (mid left centre), the arthrodire *Groenlandaspis* (bottom right) which likely fed on small worms and other creatures it could catch with its tiny needle-like teeth, and *Remigolepis walkeri* (centre right top), which may have been a bottom-feeder looking for weeds and organic-rich muds. The main predators here were the robust lobe-finned fishes called tetrapodomorphans because their skull and limb bone patterns closely resembled those of the first tetrapods, or four-legged land animals. The large *Mandageria fairfaxi*, close to 2 m long, is shown here catching an unwary lungfish *Soederbergia*, a long-snouted form also known from Greenland and North America.

Emerging from below the leaf litter (bottom left) is the small-eyed robust fish called *Gooloogongia loomesi*, an early rhizodontid. These were the giants of the early bony fish world whose later forms like *Rhizodus* would reach 6 m long in the early Carboniferous Period. *Gooloogongia* was one of the most primitive of all known rhizodontids and, together with much older fragmentary remains of the group from Antarctica, is support for a Gondwanan origin for the group. Its slender projecting lower jaw teeth and flat head shape made it an opportune lunge ambush predator that probably surprised its prey by rising up from the murky waters like a crocodile. Similar to crocs, the limbs had a rounded shoulder joint on the humerus, suggesting these large fishes were able to twist their prey in a death roll that both killed and tore them apart.

The fish swimming centre right is *Cabonnichthys*, another lobe-finned fish closely related to *Mandageria* but only half its size. Both of these fishes are members of the tristichopterid family, a group known throughout the world at this time. Recent studies, however, suggest that the Australian tristichopterids may represent an endemic Gondwana group only distantly related to those in the Northern Hemisphere. All the lobe-finned fishes shown here had powerful tusks in their jaws and were clearly predators. ●

AGE

Late Devonian (Famennian), about 363 million years ago.

LOCALITY

Near the township of Canowindra, New South Wales.

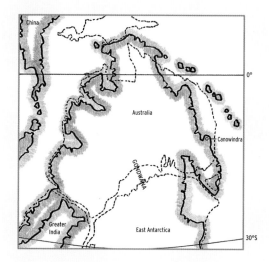

ENVIRONMENT

Fluviatile, river deposit overflow (billabong) that has dried up, leaving the fishes to die exposed. At this time Australia was further north than today, and slightly warmer with higher levels of oxygen creating lush forests near the waterways.

REFERENCES

3, 4, 191, 192, 193, 221

Mansfield, Victoria

HERE WE SEE THE LARGE RHIZODONTID FISH *Barameda decipiens* (top centre) seizing a lungfish *Delatitia breviceps* (left centre) in its powerful jaws. *Barameda*, from an Aboriginal word meaning 'fish trap', is known from two species at Mansfield. The smaller *B. mitchelli* was one of the first of the rhizodontid group to have the skull preserved and described. It grew to about 1 m long. *B. decipiens* was the largest predator in these ancient large meandering rivers and may have reached 4 m in length. It demonstrates that rhizodonts diverged early in the tetrapodomorph group. This is borne

out by the fact that the oldest and most primitive known members of the group occur in Australian and Antarctic middle Devonian deposits.

At the top right we see the sharp-spined stem shark *Gyracanthides murrayi*, which is a member of the basal shark group called acanthodians that went extinct at the end of the Palaeozoic Era. *Gyracanthides* spines are heavily ribbed and very sharp, a defence mechanism against the larger predators that shared the same rivers. A school of ray-finned fishes of the *Mansfieldiscus* variety (fish from Mansfield) are seen swimming nonchalantly away in the top right corner. These were the 'trout' of the late Devonian-early Carboniferous rivers, actively feeding on smaller fishes and invertebrates in the ancient river system.

Below is an arthrodire placoderm *Groenlandaspis*, which is thought to have died out at the end of the Devonian Period (359 million years ago) but is included here to highlight the uncertainty surrounding the true age of the Mansfield Basin deposits. This assemblage is enigmatic because it was thought to represents the very latest Devonian fauna of fishes, although some have argued it was early Carboniferous (about 350–330 million years ago) in age after all of the placoderms were gone. This scene takes the radical view that this fauna is latest Devonian, about 359 million years old, which is right on the Devonian/Carboniferous boundary. Much older remains of placoderms like *Bothriolepis* and *Austrophyllolepis* have been found in older fluviatile sedimentary rock outcroppings in the South Blue Range near Mansfield, first discovered by Australian geologist Edwin Sherbon Hills in the 1930s. Recent finds of teeth have revealed the presence of small fossil sharks in both the South Blue Range succession and in the Mansfield Basin red beds. These indicate possible proximity of these riverine communities to those of the open sea. ●

AGE

Latest Devonian-earliest Carboniferous, about 359 million years ago.

LOCALITY

Devils Plain Formation, Broken River region, Mansfield, Victoria.

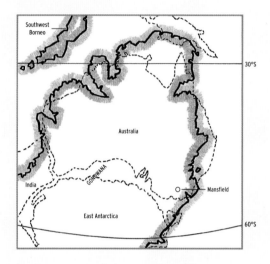

ENVIRONMENT

The fish fossils at Mansfield are all found in red mudstone and sandstone deposits indicative of large meandering river settings. At this time, Australia had a warm to tropical climate, with oxygen in the atmosphere at above today's levels. Forests made of large lycopods and ferns covered the land near the waterways.

REFERENCES

139, 180, 216, 441

Ducabrook Formation, Queensland
CARBONIFEROUS

PETER SCHOUTEN-2013

THE LOWER CARBONIFEROUS OF NORTHERN
Australia was a time of warm to arid climates, with large meandering rivers dominating the landscape. Lush forests fed by high oxygen levels in the atmosphere lined the waterways. These forests were probably inhabited by large arthropods such as occur elsewhere in the world at this time. Rotting vegetation and abundant insect life would have enriched the lake and river systems. These waterways were home to a variety of unusual fishes and one of the oldest well-preserved tetrapods (early amphibians) known from Australia.

Ossinodus pueri (lower frame), meaning 'toothed bone', was one of the most primitive tetrapods of this period, and the only Carboniferous tetrapod known from Australia. Its salamander-like body was probably covered in fish-like scales. It possessed a flat broad skull that was similar in overall form to that of the late Devonian amphibian *Acanthostega* and was around 1.5 m long.

This tetrapod is accompanied by two common fishes found in the same beds. A large spined stem shark ('acanthodian') *Gyracanthides hawkinsi* (above centre) probably grew to about 60 cm in length and may have fed on plants and small invertebrates. The large rhizodontid sarcopterygian *Strepsodus* sp. (top right) was one of the largest bony fishes (osteichthyans) of this period, with some Scottish species possibly growing as large as 4 m. It is not clear whether the Queensland species, which was about 1 m in maximum length, is actually a species of *Strepsodus* or a member of a different but similar group. It is only known from fragmentary shoulder girdle bones called cleithra. These fishes were armed with long sigmoid teeth in the lower jaw and large marginal teeth on both jaws, a clear indication that they were predators. They would most likely have hunted the swift moving trout-like palaeoniscoids, early ray-finned fishes, whose remains are commonly found as isolated scales and bones in contemporaneous deposits from central Queensland. ●

AGE
Early Carboniferous, about 330 million years ago.

LOCALITY
Ducabrook Station, near Emerald, central south Queensland.

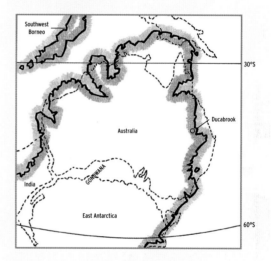

ENVIRONMENT
The fossils occur in a sequence representing a river delta, with bones deposited by a flood event. At the time, Australia was relatively warm with high oxygen levels giving rise to abundant land plants, mainly lycopods and ferns. Australia was at a similar latitude and orientation to its position today.

REFERENCES
194, 397, 405, 410

Blackwater Shale, Queensland

PERMIAN

A LARGE LAKE FORMED IN SOUTH CENTRAL
Queensland about 270 million years ago. Plant material accumulated in the lake from the surrounding forests ultimately led to the formation of black coal deposits. The lake was also home to some very interesting species of extinct fishes unique to Australia. This scene is dominated by the small shark *Surcaudalus rostratus*, known from a series of nearly complete specimens from the coal deposits at Blackwater. Its generic name derives from the well-developed epicaudal or lower lobe on the tail fin, while *rostratus* refers to the well-

developed fleshy rostrum on the head. *Surcaudalus rostratus* grew to about 20 cm long. Each of its teeth had three hooked cusps, a primitive condition seen in several much older Devonian sharks like *Phoebodus*, and an indication that it was a predator of other fishes. Both dorsal fins had sharp prominent spines, which suggests a possible relationship to the older *Ctenacanthus*.

Living alongside the shark were several kinds of ray-finned fishes. The most abundant of these was the deep-bodied form called *Ebenaqua ritchiei*, here seen swimming away in the top left of the panorama. The generic name is Latin for 'black water' and the species honours Dr Alex Ritchie. *Ebenaqua* had a very small mouth that lacked teeth. It most likely grazed on algae and other plants in the lake. The fishes shown in the centre being chased by the shark, and at the bottom left, represent two kinds of new unnamed ray-finned fishes whose fossils were described in an unpublished thesis. The school of fishes being hunted are a kind of urosthenid, a group of more advanced ray-fins also represented in the well-known Jurassic Talbragar deposits of central New South Wales. ●

AGE
Late Permian (Wuchiapingian), about 254–259 million years ago.

LOCALITY
Rangal Coal Measures, Utah Development Corporation mine, near Blackwater, southern Queensland.

ENVIRONMENT
Australia was located closer to the South Pole at this time, close to the Antarctic Circle. Climate was cool to temperature as Gondwana emerged from the early Permian ice age. The habitat for the Blackwater deposit was a freshwater coal swamp.

REFERENCES
72, 212, 217

Blina Shale, Western Australia

TWO CHARACTERISTIC EARLY AMPHIBIANS ARE shown here, the long snouted *Erythrobatrachus noonkanbahensis* to the left and the aptly named triangle-headed *Deltasaurus kimberleyensis* to the right. This scene is unusual in that it shows amphibians living in a marine environment, although it is possible that these tetrapods might have been swept out to sea before they were buried. The Blina Shale also contains remains of lingulid brachiopod shells, which also suggests the deposit formed in saltwater, perhaps an estuary. The outcrops of this fossil deposit

occur in the Erskine Ranges between Derby and Fitzroy Crossing. They were excavated by US palaeontologist John Cosgriff after fossils were first discovered by another US expert, Dr Charles Camp, on a foraging trip to the region in the late 1960s. A partial skull of *D. kimberelyensis* is known from the Blina deposit, as well as the skull of a smaller species, *D. pustulutus*, that was recovered from deep underground in a drill core near Geraldton. The Blina species grew to about 1.25 m in length. The genus is also known from skulls found in Triassic deposits in Tasmania. Recent analysis suggests that *Deltasaurus* was a member of an endemic clade of Australian rhytidosteid amphibians in the family Derwentiidae.

Erythrobatrachus belonged to the trematosaur group of extinct labyrinthodont amphibians. Growing to about 2 m in length, it may have used its long crocodilian-like snout with many needle-like teeth to catch small fish. Living alongside it in the estuarine environment was a variety of potential fish prey, including the long snouted gar-like *Saurichthys*, and medium-sized lungfish-like *Ceratodus*. The broader-headed *Deltasaurus* had more powerful jaw muscles and may well have fed on fish in the estuary, such as the ceratodontid lungfishes. ●

AGE

Early Triassic, about 247-251 million years ago.

LOCALITY

Blina Shale, Erskine Ranges, between Derby and Fitzroy Crossing, Western Australia.

ENVIRONMENT

A shallow marine seaway or estuary close to a large river mouth, where red sands and silts were deposited that buried the bony remains. At this time, northern Australia had a cool to temperate and relatively dry climate. Southern Australia was situated over the South Pole.

REFERENCES

86, 87, 99

Knocklofty Formation, Tasmania
TRIASSIC

TASMANIA AT THE BEGINNING OF THE AGE OF dinosaurs was situated close to the South Pole. Cool temperate forest filled with tree ferns and seed ferns like *Dicroidium* dominated the landscape. Here we see *Tasmaniosaurus*, a primitive thecodontid reptile, in the centre of the scene, seizing a *Banksiops townrowi* amphibian out of the river. *Tasmaniosaurus* is known from a partial skeleton found in Crisp & Gunn's quarry in West Hobart in 1960 by Max Banks and John Townrow. It was about 1 m in length and was thought to be a sister taxon to proterosuchid reptiles which were basal in the tree of archosauriforms, the group that would give rise to dinosaurs, pterosaurs and crocodiles. Recent restudy of the type material revealed that it had well-developed olfactory bulbs inside the snout. One hypothesis suggests that these organs were too large for creatures inhabiting a semi-aquatic lifestyle, hence *Tasmaniosaurus* was more likely to have been a terrestrial predator.

B. townrowi was originally thought to be a species of *Blinasaurus*, a genus first described from Triassic deposits in the Kimberley region of Western Australia, but a revision of that family indicated it was a different genus so it was given its own generic name. This creature was a typical temnospondyl amphibian about 60–80 cm in length. Its jaws were armed with sharp pointed teeth and its streamlined body shape suggests it would have been well-adapted to chase fish in the Triassic rivers that accumulated the Knocklofty Formation deposits. Potential prey would have included heavily scaled ray-finned palaeoniscoids, ceratodontid lungfishes and a coelacanth. ●

AGE

Early Triassic, about 240 million years ago.

LOCALITY

Knocklofty Formation, near Hobart, Tasmania.

ENVIRONMENT

Australia was situated over the South Pole in the early Triassic. Southern Australia was cool to cold, and forested.

REFERENCES

104, 107, 108, 293, 390, 407, 408

Arcadia Formation, Scene 1, Queensland

THE LONG-SNOUTED SLENDER RAY-FINNED FISH looking somewhat gar-like, *Saurichthys* (left centre), turns away from the oncoming predators in the river. A tiny lizard-like creature called a procolophonid, sitting on a leaf whorl of a lycophyte plant, is about to be seized by the leaping salamander-like *Xenobrachyops allos*, a small chugitosaurid amphibian. First named *Brachyops* because it was thought to be closely related to that genus found in Triassic deposits in India, it was later renamed to mean 'foreign *Brachyops*' after extensive revision of its skull

material demonstrated that it was quite unlike the Indian form. Its skull is 11 cm long, suggesting its total length was around 60–70 cm. It probably fed on the abundant small fishes and other vertebrates that shared this habitat.

A large capitosaurid amphibian *Parotosuchus gunganj* appears in the bottom left of the scene, sporting a broad flat skull with a long snout. This well-known amphibian occurred globally in the early Triassic. There are two species of this genus in the Rewan deposits of Queensland. *Parotosuchus gunganj* takes its name from the local Indigenous word meaning 'water dweller'. It grew to about 1.25 m in length and was probably an ambush hunter like many modern crocodiles. It had eyes close to the midline of the skull for viewing prey approaching the water's edge. Recent biomechanical studies of the skulls of *Parotosuchus* species reveal that the large palatal vacuities in the skull were important for decreasing skull weight and increasing the bite force of the jaws as they enabled more muscle mass to develop in the jaw region. The larger predatory amphibian *Acerastea wadeae*, which grew to 1.5 m, lurks menacingly in the background. ●

AGE
Early Triassic, about 240 million years ago.

LOCALITY
Arcadia Formation, Rewan Group, The Crater, near Rewan, Queensland.

ENVIRONMENT
Warm temperate climate, forests filled with tree ferns, seed ferns, regular ferns and progymnosperms. At this time northern Australia had a cool to temperate and relatively dry climate. Southern Australia was situated over the South Pole.

REFERENCES
183, 207, 395, 404, 406, 407

Arcadia Formation, Scene 2, Queensland

THE PIG-SIZED MAMMAL-LIKE REPTILE CALLED A dicynodont is seen here coming down to drink (centre) as a predatory reptile *Kalisuchus* lunges up out of the water to take a small unwary amphibian *Arcadia myriadens*. *Kalisuchus* had a hooked snout with large sharp teeth. Its name derives from the Hindu goddess Kali, 'the destroyer', because the original fossil material was found as a pile of hundreds of small fragments that needed to be pieced together. *Kalisuchus* grew to about 3 m in length and was one of the largest terrestrial predators in this ancient Queensland ecosystem.

The dicynodont resembles a creature called *Kannemeyeria*. These are known from South Africa where they were one of the most abundant land animals of the Triassic landscape. Dicynodonts possessed large tusks for rooting out plant material. The Australian species are only known from tusk fragments, so identification to genus is uncertain. They belong to the synapsid (mammal-like reptile) group because in terms of relationships they are closer to mammals than any other reptiles in this scene. Dicynodonts on other continents died out at the end of the Triassic but in

Queensland they persisted much longer, into early Cretaceous times.

A prolacertid reptile named *Kadimakara* is seen at the left resting on the riverbank, looking a little like an early goanna. They were lizard-like hunters that preyed on smaller reptiles, fishes and amphibians in the same area. *Kadimakara* was once thought to be a close relative of the dinosaurs and a member of the relatively derived Archosauriformes, but a recent restudy of its fossil remains places it closer to the more archaic South African reptile *Prolacerta*. ●

AGE
Early Triassic, about 240 million years ago.

LOCALITY
Arcadia Formation, Rewan Group, The Crater, near Rewan, Queensland.

ENVIRONMENT
Warm temperate climate, forests filled with tree ferns, seed ferns, regular ferns and progymnosperms. At this time northern Australia had a cool to temperate and relatively dry climate. Southern Australia was situated over the South Pole.

REFERENCES
108, 389, 392

Hanson Formation, Antarctica

ALMOST 200 MILLION YEARS AGO, THE SOUTH Pole was located just south of Antarctica which was the central hub of the supercontinent of Gondwana. The warm temperate climate here was favourable to forests full of ferns and cycads, and dinosaurs and mammal-like reptiles ruled the land. This assemblage of creatures was collected at an altitude of nearly 4000 m in the Transantarctic Mountains by Dr Bill Hammer and his team from Augustana College in the 1990s. It has yielded the oldest dinosaur fauna from that continent. The generic name of the predatory *Cryolophosaurus*

ellioti (centre right) means 'frozen lizard'. The species name honours its finder, Dr David Elliot. It was a medium-sized theropod up to 6.5 m long. It is known from much of the skeleton and a near-complete skull. It bore distinctive swept-back crests above the eye ridges, possibly serving as an enhanced visual display to attract mates. *Cryolophosaurus* is regarded as a basal tetanuran, meaning it was a very primitive kind of predatory dinosaur not related to the well-known allosaurs or tyrannosaurs of the Northern Hemisphere. Its prey may well have included the 7.5 m

Glacialisaurus hammeri (centre left). This generic name means 'icy lizard' and the species name honours Bill Hammer. *Glacialisaurus hammeri* was a massospondylid sauropodomorphan dinosaur, one of the semi-bipedal ancestors of the gigantic, long-necked forms. It fed primarily on ferns and cycad plants. It is closely related to forms in China like *Lufengosaurus*. The small weasel-like creature (bottom left) is a tritylodont, a kind of advanced mammal-like reptile about 30 cm in length. It was probably omnivorous, eating worms, insects, frogs and maybe some plants. It has not yet been named because it is known from only one postcanine tooth which is indistinguishable from species of *Bienotheroides* from China. ●

AGE

Early Jurassic, about 195–187 million years ago.

LOCALITY

Mt Kirkpatrick, Transantarctic Ranges, about 650 km from the South Pole, Antarctica.

ENVIRONMENT

The fossil site was located between 55°S and 65°S. Here, during the early Jurassic, Antarctica was thought to have had a warm temperate climate, as shown by the abundant plants and fossil wood which indicates that large gymnosperm trees were present in the ecosystem.

REFERENCES

153, 154, 360, 362, 363

Talbragar, New South Wales

BY THE LATE JURASSIC WHEN VERY LARGE dinosaurs were roaming the cool Australian landscape, aspects of the modern fish fauna had begun to appear in our rivers, seas and lakes. The Talbragar fish beds have yielded beautifully preserved complete specimens which show for the first time in Australian history a dominance of teleostean fishes, the group to which 99% of the modern living bony fishes belong. The scene is dominated by the presence of *Uabryichthys latus*, a large predatory macroseminonitid fish coming in from the top right, chasing a school of

smaller herring-like *Cavenderichthys talbragarensis*. They are close relatives of a group that includes *Leptolepis koonwarri* from Victoria and *Luisella* from Argentina, a clade of fishes that were endemic to Gondwana. Below them, resting on the weedy river floor, is an elongate *Archaeomene*, looking wary at the presence of the approaching predator above it.

Madarsicus is seen at the top left. Fishes with such deep body shapes mostly fed on algae and other plants in the river system. Their large size and schooling habit protected them from most other predators. However, sharks up to 1 m long are known to have existed in these rivers and would have been the top predators in the fluvial habitat.

The floor of the river shows leaves of early conifers including branches from a species of *Wollemia*. The Wollemi Pine is a modern species of this genus that survives in the northern Blue Mountains of New South Wales and has been nicknamed the 'Pinosaur'. Fossils of many different kinds of edible insects, spiders and other small invertebrates in the Talbragar deposits indicate that there was an abundance of food items available to the fishes in these waters. ●

AGE

Late Jurassic, 151 million years ago.

LOCALITY

Purlawaugh Formation, Talbragar River, near Gulgong, New South Wales.

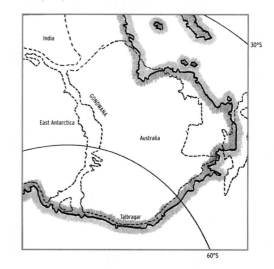

ENVIRONMENT

Australia was at a relatively high, cool temperature latitude in the mid to late Jurassic. The flora changed from an araucarian abundance to a podocarp-dominated phase.

REFERENCES

36, 352, 396, 402, 403, 440

Evolution of the tetrapod forelimb

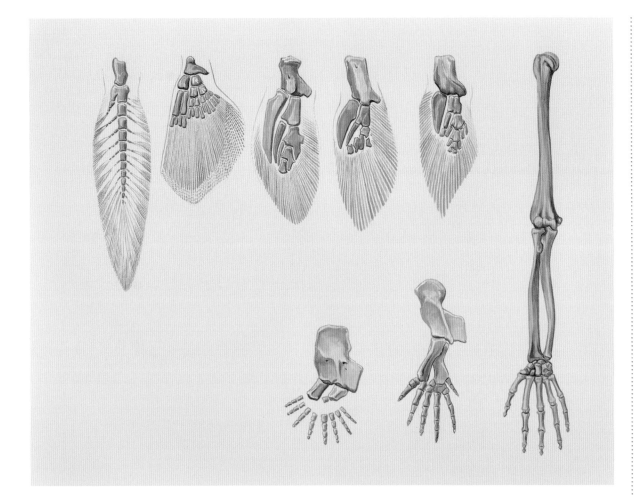

TODAY ALL LIVING LAND VERTEBRATES,
including some that have secondarily gone back into the seas, are called tetrapods, meaning they share a pattern of four limbs. Some that have secondarily gone back to life in water (e.g. whales) and others that have lost limbs (e.g. snakes) are still referred to as tetrapods because their earliest members retained the four-limbed body pattern. We call these limbs arms and legs, but they all evolved directly from the paired pectoral and pelvic limbs of fishes. This sequence shows the evolutionary stages from a primitive lobe-finned fish, a lungfish (top left) through to the human arm (extreme right).

The living Queensland lungfish *Neoceratodus* (top left) shows an archipterygium, or series of bones forming the forelimb. The first bone is an adult fused humerus and radius, these two bones being separate in embryonic lungfish. The differentiation of the ulna and radius was first established in *Onychodus* (from p. 34, Gogo), well before the first tetrapodomorphan fishes evolved.

Next along to the right are series of forelimbs belonging to tetrapod-like fishes (tetrapodomorphans) all of late Devonian age (about 380–359 million years ago). The forelimb of *Sauripterus* (a rhizodontid) has a robust humerus and a well-defined ulna and radius. Third from the left is *Gogonasus*, from Gogo. Here the humerus is getting larger and the ulna more prominent. The next two to the right are *Panderichthys* (fourth from left) and *Tiktaalik* (fifth from left); both are elpistosteglians, which are advanced lobe-finned fishes that were very close to becoming tetrapods. Both have much larger and more robust humerus–ulna–radius patterns with the appearance of bony digits in the wrist area, which were soon to descend and form the bone supporting the hand. *Tiktaalik* has an incipient wrist joint. In *Sauripterus* we see a robust broad humerus with a long entepicondylar process. This became relatively shortened in more progressive forms like *Gogonasus* and *Panderichthys*, which were more closely related to the first tetrapods. In *Acanthostega* this process is almost square-shaped. It is broad and flat for attachment of the lower arm muscles.

The lower series of illustrations shows the limbs of very early tetrapods. *Acanthostega* (bottom left) and *Tulerpeton* (bottom centre), both of late Devonian age, show a direct comparison of the major limb bones with that of a human arm (right). This demonstrates the continuity throughout evolution of a persistent pattern once it became established.

Digits seem to have first appeared within the fin of tetrapodamorph fishes like *Panderichthys* and *Tiktaalik*, although they were incorporated into the wrist area. Well-defined digits splaying out as a hand or foot first evolved in early tetrapods like *Acanthostega*, which had eight digits on its limbs, and *Tulerpeton*, whose limbs bore six digits. The oldest known tetrapods with the pentadactyl system of five fingers and toes is *Pederpes* from the early Carboniferous of Scotland. The human arm shows the same pattern of bones as the Devonian fish. The additional digits forming the hand were apparent in early amphibians like *Tulerpeton*, but the specialised bones of the wrist are a strictly reptilian feature that mammals inherited. ●

AGE
Devonian to Recent, about 380 million years ago–today.

REFERENCES
57, 75, 78, 209, 228, 356

Broome Sandstone, Western Australia

DINOSAUR TRACKS HAVE BEEN KNOWN FROM the Broome Sandstone outcrops by the local Indigenous Bardi people for thousands of years. They believe that Marala, the Emu Man, left his footprints there along what is now called the Lurujarri Heritage Trail. In 1968 US palaeontologist Ned Colbert, working with Duncan Merrilees of the Western Australian Museum, published the first paper describing theropod trackways at Minyirr (Gantheaume Point), which they named *Megalosauropus broomensis*. These tracks were made by a medium-sized dinosaur, perhaps 4 m in length. Tony Thulborn continued investigations of these trackways and noted that some, which were 1.75 m in length, must have been made by an enormous sauropod (the one illustrated here as a probable titanosauromorph). The staggering size of these footprints indicate it was probably the largest dinosaur known from anywhere in the world. More recent investigations by Steven Salisbury and his team from the University of Queensland, working closely with local Indigenous Elders and other local volunteers, have resulted in recognition of at least 21 different kinds of dinosaur tracks – the most diverse assemblage of dinosaur trackways of this age anywhere in the world. In addition to the giant sauropod tracks noted above, others such as the ones named *Oobardijama foulksi* (little thunder, in Nyulnyulan language) represent more modest-sized animals around 12 m long. These trackways have also produced the first record of Cretaceous armoured dinosaurs (thyreophorans) like those named *Garbina roeorum*, a *Stegosaurus*-like animal up to 6 m long (shown here behind the large sauropod). Several kinds of plant-eating ornithopods left tracks here too (like the one shown centre left, in background). The largest ornithopod tracks are of *Walmadanyichnus hunteri*, whose footprints measure up to 80 cm. At this size, it would have been a beast slightly larger than *Muttaburrasaurus* from Queensland. ●

AGE

Early Cretaceous, Valangian-Barremian, about 140-127 million years ago.

LOCALITY

The coastal region around Broome known as the Yanijarri-Lurujarri Heritage Trail, Western Australia.

ENVIRONMENT

At this time, northern Australia would have had a cool to temperate but relatively dry climate that nevertheless supported forests filled with tree ferns, seed ferns, regular ferns and progymnosperms.

REFERENCES

79, 220, 344, 391

Strzelecki, Victoria

ALTHOUGH THE FIRST DINOSAUR BONE, A theropod hand claw, was found in an area south of Wonthaggi, Victoria in 1904, it wasn't until the 1970s that new finds were made in the area, involving rich discoveries of seams full of fish, reptile and mammal bones. The scene here depicts some of the most spectacular early finds from the excavations at the Flat Rocks site near Inverloch. Back 120 million years ago this was a large temperate rift valley that formed as Antarctica began breaking away from Australia. In the scene we see the head of a large abelisaurid

theropod, perhaps around 6 m long, attacking the 4 m newt-like labyrinthodont amphibian *Koolasuchus cleelandi*, one of the last surviving members of this gigantic amphibian clan. *Koolasuchus* lived in large rivers and captured its prey as an ambush predator, like crocodiles do today. It was named after Leslie Kool and Mike Cleeland who helped discover and excavate its remains. To the right a ceratodontid lungfish, the 1 m *Archaeoceratodus avus*, is swimming way from the violent confrontation. This fish is only known from a broken upper tooth plate and a scale. It has been argued to be an ancestor on the line leading to the modern Queensland lungfish. The theropod head shown here was originally identified as belonging to the well-known North American dinosaur *Allosaurus*, based on discovery of a single ankle bone (astragalus). This was later referred to the megaraptoroid *Australovenator*, known from Winton in Queensland. Dinosaur experts working in South America have argued that it probably belonged to an abelisauroid dinosaur, a group of deep-headed short-armed predators known mainly from South America and Madagascar. This is quite likely to be correct, given that at this time Australia was closer to South America which at the time was an integral part of Gondwana along with Antarctica. ●

AGE
Early Cretaceous, Aptian stage, about 120 million years ago.

LOCALITY
The coastal shore platforms from Cape Paterson through to Inverloch, eastern Victoria.

ENVIRONMENT
At this time, southern Victoria had a cool temperate climate which supported forests filled with tree ferns, seed ferns, regular ferns and progymnosperms. Southern Australia lay within the Antarctic Circle.

REFERENCES
1, 203, 328, 330, 409

Bulldog Shale, South Australia

FOSSILISED BONES PRESERVED AS OPAL ARE
very rare. They are only found in abundance in
Australia from the opal mining regions around
Coober Pedy in South Australia and Lightning Ridge
in New South Wales. At this time a vast sea covered
inland Australia, home to a diverse range of marine
life including huge clam-like molluscs, squid-like
belemnites, coiled shell ammonites, many kinds
of fishes, sharks and marine reptiles. In this scene
we see *Umoonasaurus desmoscyllus*, a 2.5 m
pliosaur (short-necked plesiosaurian), chasing

a small ray-finned fish, similar to *Boreosomus*
that lived in the Triassic Period in the Northern
Hemisphere. *Umoonasaurus* was discovered by
opal mining machinery, hence the skeleton was
originally crushed into hundreds of small pieces.
It was painstakingly prepared and glued together
over 450 hours by Dr Paul Willis, who nicknamed
it 'Eric' after the Monty Python song 'Eric the
Half Bee'. This near-whole skeleton is the most
complete of the known opalised skeletons found
in Australia. Its skull shows the animal had well-

developed crests over the eyes and snout, likely covered by keratin extensions in life, perhaps for communication or sexual displays. Alternatively, the crests might have aided the animal in steering through the water when chasing down its fishy prey. Phylogenetic analyses show it is closely related to either the leptocleidid or polycotylid plesiosaurs. *Umoonasaurus* is named after the local Antakirinja language and means 'lizard from Coober Pedy region'. The species name means 'sea monster of the people', which honours the fact that the specimen – made of valuable opal – had to be bought from its owners to join the collections of the Australian Museum. This was achieved by a successful public fund-raising campaign to heighten awareness of its importance. Another skeleton of this beast, that of a juvenile, is held in the South Australian Museum. ●

AGE

Early Cretaceous, Aptian-Albian stages, about 115 million years ago.

LOCALITY

Coober Pedy, central South Australia.

ENVIRONMENT

The presence of ice-rafted boulders and glendonites representing ikaite, a mineral that forms only between -1°C and 6°C, in the same deposit suggest a temperature range that was seasonal, with cool to very cold winters that possibly involved surface water freezing at times.

REFERENCES

41, 110, 198, 201, 202

Otway, Scene 1, Victoria

IN THE COOL POLAR FORESTS OF SOUTHERN
Victoria 106 million years ago, a spring scene of
primeval tension unfolds. The 2.5 m predator
Timimus hermani (top left) spoils the moment
when the similarly sized feathered megaraptorid
dinosaur was about to pounce upon the small
ornithopod *Leaellynasaura* emerging after its
winter hibernation from its burrow in the riverbank.
Timimus (named for both Tim Rich and Tim
Flannery, and the species for John Herman) was at
first thought to belong to one of the ostrich-like
ornithomimid dinosaurs known only from Northern

Hemisphere countries but some workers now
assign it to the tyrannosauroid group (as depicted
here), while others regard its affinities among
theropods as uncertain. It is only known from
its slender leg bones, two femora found within
1 m of each other. The unnamed megaraptorid
shown here is based on a solitary arm bone (ulna)
from the same site which has features close
to those of the large *Megaraptor* from South
America. These dinosaurs had elongated arms
with sickle-shaped claws for grasping their prey.
The two turtles depicted are forms known from

reasonably well-preserved carapaces and isolated bones. *Chelycarapookus arcuatus* (bottom left, underwater) was found at a road cutting near Carapook in western Victoria. It is thought to be related to the modern long-necked turtles. *Otwayemys cunicularis* (bottom right, on land) is a cryptodire, the group containing most living tortoises and turtles. Its closest fossil relatives are from the Jurassic of China, suggesting it was a basal member of the group. ●

AGE

Early Cretaceous, Albian stage, about 106 million years ago.

LOCALITY

Eumeralla Formation outcrops and subsurface strata, Dinosaur Cove and coastal cliffs near Apollo Bay, western Victoria.

ENVIRONMENT

Cool temperate climate, forests filled with tree ferns, seed ferns, regular ferns and progymnosperms. Southern Australia lay within the Antarctic Circle.

REFERENCES

42, 95, 135, 181, 329, 361

Otway, Scene 2, Victoria

LEAELLYNASAURA AMICAGRAPHICA WAS THE FIRST dinosaur to be formally named from Victoria, in 1989, based on discoveries from Flat Rocks and other sites near Inverloch. Its name is after Leaellyn Rich, the young daughter of palaeontologists Tom and Patricia Vickers-Rich who discovered and described the fossils, aided by a large team of volunteer helpers. Its species name honours the National Geographic Society for sponsorship of the work. This dinosaur was only about 90 cm in length. It was a basal ornithopod, the main clade of plant-eating dinosaurs. Because it lived in a cool climate close to the South Pole, it would have endured harsh dark winters. Its skull revealed a large margin for the eye socket and impressions of the brain showed it had enlarged optic lobes, suggesting it had large eyes with acute vision. Its teeth have fine striations perfect for slicing up ferns and other rainforest plants that were abundant in its habitat. It was a member of the basal ornithischian group, similar to hypsilophodontids. Several other genera of small plant-eating dinosaurs have been found alongside *Leaellynasaura*, comprising the commonest group of dinosaurs in this assemblage.

The two possible reconstructions of *Leaellynasaura* shown here include a conventional naked one with ornithopod-like reptilian skin (above) and a version covered with hair-like feathers or quills (below). The latter is based on the fact that some theropod dinosaurs living in China at this time developed feathers or quill-like structures, as did some ornithischians like *Tuanylong*. ●

AGE

Early Cretaceous, Albian stage, about 106 million years ago.

LOCALITY

Coastal outcrops near Inverloch, eastern Victoria.

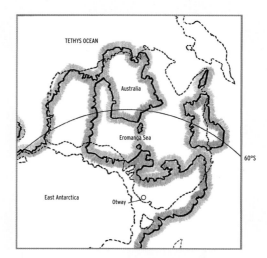

ENVIRONMENT

In the cool to cold temperate climate, the forests were filled with tree ferns, seed ferns, regular ferns and progymnosperms. Because southern Australia lay within the Antarctic Circle, it would have been dark for three months of the year.

REFERENCES

171, 172, 330

Toolebuc Formation, Scene 1, Queensland

CENTRAL NORTHERN QUEENSLAND WAS covered by a vast inland sea about 100 million years ago. Some of the first and most spectacular finds from the Toolebuc Formation, the result of sediments accumulating in an expansive inland sea in the area of central northern Queensland, include the gigantic pliosaur *Kronosaurus queenslandicus*. First recognised and named from a fragment of the snout by Heber Longman of the Queensland Museum in 1924, a near-complete skeleton was retrieved in the 1930s by a Harvard University expedition in 1931. Australian museums were invited to join the expedition but were unable to participate, hence the skeleton was sent to the US and prepared over 25 years before finally going on display in Harvard University's Museum of Comparative Anatomy. Since then, many new specimens have been uncovered in the Winton–Richmond–Hughenden area of Queensland, and some are on display in the museums of these towns. The new discoveries demonstrate that a crest included at the back of the skull of the Harvard specimen was actually a mistaken plaster addition – the animals had no crest. When this was realised, for a short while the Harvard skeleton earned the nickname 'Plasterosaurus'. *Kronosaurus* is named after Cronos, the Greek Titan, who ate his children. Growing up to 11 m in length, it was the most ferocious predator in the seas around Australia at this time. Its powerful jaws sported elongated teeth up to 30 cm long. These lack carinae (distinct cutting edges) so are easily distinguished from the teeth of other giant pliosaurs like the British *Liopleurodon*. Stomach contents of *Kronosaurus* showed it preyed upon turtles and long-necked plesiosaurs like *Eromangasaurus*, shown here being attacked. A phylogenetic analysis shows *Kronosaurus* was perhaps the most specialised member of its lineage of gigantic pliosaurians. ●

AGE
Early Cretaceous, late Albian, about 106–103 million years ago.

LOCALITY
The region around the towns of Winton, Richmond and Hughenden in central Queensland.

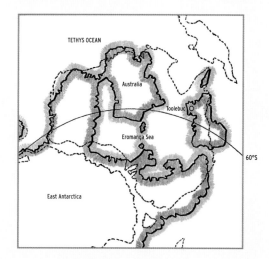

ENVIRONMENT
These marine reptiles lived in a cool to temperate shallow inland sea. At this time northern Australia had a relatively dry climate.

REFERENCES
205, 229, 242, 340

Toolebuc Formation, Scene 2, Queensland

CRETACEOUS

CENTRAL NORTHERN QUEENSLAND WAS covered by a vast shallow inland sea around 100 million years ago. Ichthyosaurs like *Platypterygius longmani* shown here (centre top) were fully aquatic dolphin-like reptiles with long snouts. This oceanic fish-eating predator was Australia's only Cretaceous ichthyosaur. It is known from several complete skeletons which indicate that it grew to around 7 m in length and was armed with conical stout teeth perfectly suited for catching fish. The long-necked plesiosaur in the bottom left of the scene is *Eromangasaurus australis*, an elasmosaurid,

all members of which have much longer necks than other plesiosaurians. Its long needle-like teeth suggest it hunted relatively small fish, as did the large predatory ichthyodectiform fish *Pachyrhizodus marathonesis* (seen above the plesiosaur, centre left). The saw shark *Pristiophorus tumidens* (bottom right) is lying on the sea floor. Like modern saw sharks, it would have slashed its deadly toothed rostrum into schools of fishes in order to feed on the wounded ones. It is only known from this site by its fossilised teeth. These seas were also home to a variety of ancient turtles. The little *Notochelone costata* (centre top) was about 1 m long and is represented by complete carapace fossils. It might have lived much like the Green Turtle today. *Cratochelone berneyi* (top right, background) is known from bones of the shoulder girdle. Their large size suggests that this turtle grew to about 2 m in length. Like modern turtles, they would have been omnivores feeding on fish, crabs, sponges, shellfish, jellyfishes, algae and sea grasses. ●

AGE

Early Cretaceous, late Albian, about 106–103 million years ago.

LOCALITY

A range of sites around the Winton-Hughenden-Richmond region of central Queensland.

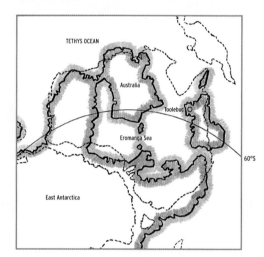

ENVIRONMENT

These marine reptiles lived in a cool to temperate shallow inland sea. At this time northern Australia had a relatively dry climate.

REFERENCES

197, 198, 199, 342

Toolebuc Formation, Scene 3, Queensland

THE CARCASSES OF DINOSAURS AND pterosaurs that inhabited the land close to the Eromanga Sea of Queensland would sometimes get swept out to sea after flooding events, becoming buried in the shallow marine sediments to become part of the Toolebuc Formation or Allaru Mudstones. Some, like the armoured ankylosaur *Kunburrasaurus ieversi* (left) are known from a nearly complete skeleton that includes well-preserved body armour and a perfectly preserved skull (this taxon was previously referred to as *Minmi*). Its name means 'shield-lizard' in the

Mayi language of the local Wanumara people. At almost 3 m long, it had a well-armoured body covered by many small bony osteoderms set into its skin, and a thick bony skull. We know it fed on seeds, leaves and fungi, as evidenced by stomach contents preserved among the skeletal remains. A few vertebrae of the large, long-necked sauropod *Austrosaurus mckillopi* (right), the generic name meaning 'southern lizard', were first found in 1933 but new specimens were collected from the site on Clutha Station in 2014 by Steve Poropat and Tim Holland. It is now represented by six vertebrae and several ribs. These remains suggest it was about 4 m tall at the shoulder and up to 20 m in length. It was one of the titanosauriform sauropods, a group common in many areas of Gondwana at this time. The pterosaur *Mythunga camara* is seen here at the centre of the scene. The generic name is based on the local Indigenous name for the constellation Orion. *Mythunga* is represented by several bones of the skull and jaws, making it Australia's most complete pterosaur. It had a wingspan up to 4.7 m, making it Australia's largest-known flying animal. The sharp teeth of *Mythunga* suggest it fed on fish. In this scene, resting on the flank of *Austrosaurus* are three small *Nanatius eos*, a bird which was about the size of a modern blackbird. This bird is known from only a few tiny bones but is significant in being the only named bird from the age of dinosaurs in Australia. ●

AGE

Early Cretaceous, late Albian, about 106–103 million years ago.

LOCALITY

Richmond-Winton-Hughenden-Boulia region, central northern Queensland.

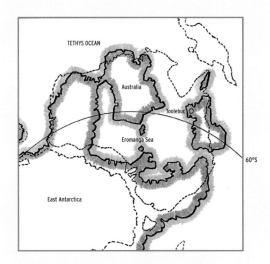

ENVIRONMENT

Warm, temperate climate, forests filled with tree ferns, seed ferns, regular ferns and progymnosperms. At this time northern Australia had a cool to temperate, relatively dry climate.

REFERENCES

208, 255, 257, 260, 291, 307

Winton Formation, Scene 1, Queensland
CRETACEOUS

A PAIR OF PREDATORY THEROPOD DINOSAURS
Australovenator wintonensis run through warm temperate forest, trampling ferns underfoot as they hiss at each other. These dinosaurs are represented by articulated and relatively complete skeletons, the first such well-preserved theropod dinosaurs known from Australia. They were first discovered in sediments near Winton, Queensland. Recent research has led to a detailed reconstruction of their foot anatomy which has enabled estimation of the degree of pedal mobility. It is now clear that they had the ability to use the middle toe as a retractable claw on each foot, a feature found in most of the raptor group. *Australovenator*, meaning 'southern hunter', was about 2 m tall at the hip and up to 6 m in length, which is large for an Australian Cretaceous dinosaur. Apart from the few large sauropods known (featured in the next illustration), most of Australia's dinosaurs of this age, such as the hypsilophodontids, were much smaller. These smaller plant-eating dinosaurs, together with lungfishes and crocodiles that lived in the streams, could have been prey items for *Australovenator*.

We know little about the lifestyle of this beast apart from what is hypothesised for other similarly sized theropods found elsewhere. Accumulations of several dromaeosaur skeletons together in the same site in the US have led to suggestions that they might have worked in packs to bring down large prey. While this may also have been the case in Australia, there is at present no evidence that this was the case. It is equally likely they were solitary hunters. Dinosaur community sizes at this time in Australia seem to have been far less dense than those in North America or Asia. The most recent phylogenetic analyses place *Australovenator* as a megaraptorid theropod, a group that probably originated in Australia before spreading to other regions of Gondwana. Their closest relatives include theropods from Europe and South America like *Neovenator* and *Chilantaisaurus*. ●

AGE

Late Cretaceous, Cenomanian Stage, about 100–94 million years ago.

LOCALITY

Winton Formation, near Winton, central Queensland.

ENVIRONMENT

While temperatures in southern Australia were very cold at this time, with polar conditions dominating, further north in Queensland was likely to have been warmer or even arid at times as suggested by the extensive red-bed deposits of sandstone and shale in the Winton Formation. Research has determined that the climate was humid with seasonal rainfall, warm to hot summers and occasional frosts in the winter (ref. 123).

REFERENCES

1, 123, 176, 259, 420, 421, 422

Winton Formation, Scene 2, Queensland

PETER SCHOUTEN · 2011

IN THE ESTUARIES, RIVERS AND BILLABONGS
that drained into the inland Eromanga Sea there
lived a variety of fishes and reptiles, including the
small crocodile *Isisfordia duncani*, which was just
over 1 m long. The land at this time swarmed with
large dinosaurs, including predatory theropods.
Hence, little crocodiles like *Isisfordia* had to stay
close to the water's edge to make a ready escape.
No ancient relict, this was a beast well ahead of its
time in that it was perhaps the earliest member
of the modern crocodilians in the group Eusuchia.
The perfectly preserved, uncrushed skeleton was
recovered from near the town of Isisford in central
northern Queensland. The species name honours
Ian Duncan who found the specimen. Its vertebrae
exhibit a slightly convex rear face, suggesting the
link to modern crocodiles which have vertebrae
with strongly convex rear surfaces (the 'procoelous'
condition). Its skull has the pterygoid bones
incorporated into its secondary palate. These
are advanced features that place it as an early
member of the eusuchian group. While *Isisfordia*
was only tiny, by the end of the Cretaceous the
next wave of eusuchians grew to enormous sizes,

some reaching nearly 10 m in length. The huge fish swimming below would have likely been too big for *Isisfordia* to tackle. It is one of the earliest members of the elopomorph group, which includes tarpons, and although not found in the exact same beds as *Isisfordia* it lived close by in the shallow marine-estuarine conditions that were widespread at that time. ●

AGE
Late Cretaceous, Cenomanian Stage, about 100-94 million years ago.

LOCALITY
Winton Formation, near Isisford, central Queensland.

ENVIRONMENT
While temperatures in southern Australia were cold at this time, with polar conditions dominating, further north in Queensland there probably were warmer or even arid conditions at times.

REFERENCES
43, 123, 166, 176, 177, 220, 306, 308, 343

Winton Formation, Scene 3, Queensland

CRETACEOUS

UNTIL RECENTLY, AUSTRALIA'S RECORD OF THE long-necked sauropod dinosaurs from the Cretaceous Period was relatively poor. However, discovery in the early 2000s of the relatively well-preserved sauropods from the Winton Formation has demonstrated that large sauropods, some up to 30 m long, lived in Queensland about 95 million years ago. *Diamantinasaurus matildae* (right) was discovered in 2005 as a partial arm bone (humerus) on Elderslie Station, north of Winton. Further remains of this sauropod were discovered between 2006 and 2009. Its name honours the Diamantina

River and the famous song 'Waltzing Matilda', written by Banjo Paterson while staying at a cattle/sheep station near Winton. *Diamantinasaurus* is now known from several bones including a partial skull and braincase. It was about 15–18 m long and about 2.5 m high at the hip. *Wintonotitan wattsi* (left) was discovered as partial remains in 1974 on Elderslie Station, but the partially complete skeleton now known was not recovered until excavations were completed in 2006. Its species name honours its finder, Keith Watts. It is now known from the front limbs, shoulder bones, body and tail vertebrae and part of the hip. In life, this dinosaur would have measured 16–20 m in length. The relationships of *Wintonotitan* are unclear but it appears to be a relatively primitive sauropod and a non-titanosaurian, in contrast to *Diamantinasaurus*. Subsequently a new sauropod, *Savannasaurus elliotorum*, was described from Winton. An analysis of the distribution of titanosaurians suggests these Australian late Cretaceous species were some of the last surviving members of a group that was once widespread across the globe during the early Cretaceous. Phylogenetic analyses suggest that *Savannasaurus* and *Diamantinasaurus* are closely related species, a little more primitive than the clade containing the lithostrotian titanosaurs – a group named because several species had solid bony plates set in the skin for protection against predators. An even larger sauropod recently found in central Queensland, named *Australotitan cooperensis*, may have been 6.5 m high at the hip and 30 m long. While *Australotitan* is the largest dinosaur known from skeletal elements, the giant footprints from Broome (p. 62) indicate that even larger dinosaurs – in fact the largest in the world – are still waiting patiently to be discovered as skeletons. A small dolichosaur (bottom right), similar to late Cretaceous species of the Northern Hemispheric genus Coniasaurus, watches carefully to avoid the heavy feet of the giant sauropods. These aquatic ophidiomorhan lizards may have been very close relatives of the gigantic marine mosasaurs. ●

AGE

Late Cretaceous, late Albian, about 100–94 million years ago.

LOCALITY

Winton Formation, near Winton, central Queensland.

ENVIRONMENT

While temperatures in southern Australia were cold at this time, with polar conditions dominating, further north in Queensland there probably were warmer and even arid conditions at times.

REFERENCES

123, 304, 306, 308

Lightning Ridge, New South Wales

ONE OF THE MOST FAMOUS OPAL MINING AREAS in the world is Lightning Ridge, New South Wales, which has produced some of the world's most extraordinarily beautiful black opal gemstones. What is less widely known is that many of these gems were in fact fossils of Cretaceous animals and plants. How many unique species have been ground down over the last century to make jewellery may never be known. Fortunately, some very important fossil specimens have survived. The animals and plants represented by these specimens lived in or were washed into large

freshwater lagoons at a time when much of Australia was covered by vast inland seas. Among fossils that did not get converted into gemstones are the teeth of lungfish including *Neoceratodus potkooroki*. Ferocious theropod dinosaurs such as *Rapator ornitholestoides* (the feet sloshing here through the water) would have terrorised smaller dinosaurs in the adjacent forests. An archaic egg-laying monotreme *Steropodon galmani* (floating near the surface) was a distant cousin of ornithorhychid platypuses but, because of differences in tooth and jaw morphology, has been allocated to its own unique family, the Steropodontidae. The name *Steropodon* comes from two Greek words meaning 'lightning tooth' in reference to the locality as well as the flash of colours in the specimen. A much weirder mammal that appears to have been a highly specialised monotreme with molar teeth that look like hot-cross Easter buns was *Kollikodon ritchiei* (here shown crunching a mussel shell it has foraged from the bottom mud). It too has been allocated to its own family, the Kollikodontidae. Because of its bizarre tooth form, it was given the informal nickname of Hotcrossbunidon. While most of the animals and plants found in the Lightning Ridge fossil deposits were terrestrial or at least freshwater occupants, such as pinecones and turtles, occasional discoveries of plesiosaur and shark teeth indicate that the lagoon must sometimes have been in direct contact with the adjacent inland sea. ●

AGE

Late Cretaceous, Cenomanian, about 100 million years ago.

LOCALITY

Griman Creek Formation, Lightning Ridge, New South Wales.

ENVIRONMENT

While most of the fossils from this formation are freshwater invertebrates, the diversity of land vertebrates and plant fossils indicate that the adjacent land was almost certainly dominated by species-rich forests. Given its geographic position – more or less between the areas where the Cretaceous Winton Formation in Queensland and Eumeralla Formation in Victoria deposits accumulated – the climate may have been intermediate between those of these two areas (warm and cool respectively), and hence perhaps mild.

REFERENCES

19, 40, 119, 204, 292

Mackunda Formation, Queensland

CRETACEOUS

MUTTABURRASAURUS ('THE LIZARD FROM Muttaburra') was one of the first relatively complete large dinosaurs found in Australia, and certainly the first large plant-eating ornithopod dinosaur known from this country. It was discovered in 1963 on Muttaburra Station in the Mackunda Formation, a marine deposit in central Queensland. The discoverer, grazier David Langdon, alerted scientists at the Queensland Museum who excavated the bones. The specimen took many years to prepare using fine chisels and a range of different acids. It was finally published in 1981 by Alan Bartholomai and Ralph Molnar.

It was a bizarre-looking beast about 10 m in length, that had a strangely expanded part of the skull associated with the underlying nasal chamber. The nasal chamber might have enabled it to make unusual calls, or perhaps increased its sense of smell. It was interpreted to be an iguanodontid, a dinosaur that had a spike-thumb. Another specimen found near Dunluce, Queensland, from the slightly older Alluru Mudstone exhibits just such a structure. Other bones attributed to *Muttaburrasaurus* found in Lightning Ridge, New South Wales, suggest that more than one species might be present. The most recent analyses of its features suggest *Muttaburrasaurus* is in fact a very basal ornithopod most closely related to dinosaurs from Transylvania like *Rhabdodon*. This would suggest it may be a rhabdontomorphan rather than an iguanodontid dinosaur.

In the scene we see a young *Muttaburrasaurus* on all fours with a detail of the head of a second animal in the foreground, showing the unusual nasal swelling. The reconstructed skeleton of *Muttaburrasaurus* stands proudly in the entrance hall of the Queensland Museum in Brisbane, mounted as if standing on its hind legs. We now know, however, from trackways and biomechanical analyses, that although these dinosaurs were capable of stretching up with their arms to reach branches, they were mostly quadrupeds.

The teeth of *Muttaburrasaurus* have several strong ridges and are unusual in that they erupted all at the same time, unlike other ornithopods whose teeth erupted randomly. It was even suggested by the original authors that, although well-adapted for chewing hard ferns and conifer leaves, *Muttaburrasaurus* may have been an omnivore that scavenged carcasses from time to time. ●

AGE

Late Cretaceous, about 98-95 million years ago.

LOCALITY

Mackunda Formation, Muttaburra Station, east of Winton, central Queensland.

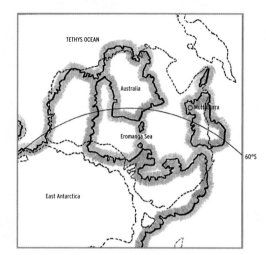

ENVIRONMENT

While temperatures in southern Australia were cold at this time, with polar conditions dominating, further north in Queensland there were probably warmer and even arid conditions at times.

REFERENCES

35, 220, 237, 256

Mangahouanga, New Zealand
CRETACEOUS

IN THE CRETACEOUS SEAS AROUND WHAT WAS
to become New Zealand lived gigantic marine
reptiles similar to those that ruled the world's
oceans at this time. In the North Island of New
Zealand, the Tahora Formation exposed by
the Mangahouanga Stream, a tributary of the
Te Hoe River, has produced two such reptiles.
Large long-necked elasmosaurs described as
Tuarangisaurus keyesi, family Elasmosauridae,
paddle about on the surface and are being eyed
by a mosasaur. These elasmosaurs were predators
of smaller animals such as fish and cephalopods.
However, *Moanasaurus mangahouangae*, family
Mosasauridae, was the top predator in these
Cretaceous oceans. It grew to about 12 m in length
with a skull that was about 75 cm long. It would
almost certainly have attacked elasmosaurs
from time to time. Both of these species were
discovered by Joan Wiffen and colleagues. Wiffen
was a remarkable amateur palaeontologist who
went in search of dinosaurs and, while doing so,
found these huge marine reptiles. She named the

first of these taxa in 1980, when she was 58 years old, and continued fieldwork in the rugged hills and rivers for the next 20 years. She carried out the fossil-bearing rocks from the site and prepared them in her backyard. Her hard work over many years has resulted in the recovery of some truly spectacular fossils and significantly expanded understanding about the distinctive Mesozoic marine reptiles of New Zealand. ●

AGE

Late Cretaceous (Campanian-early Maastrichtian), about 73 million years ago.

LOCALITY

Mangahouanga Stream, Hawke's Bay, New Zealand.

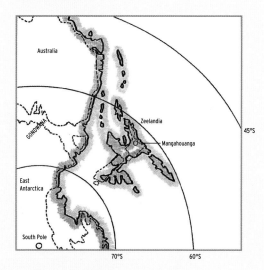

ENVIRONMENT

These giant sea dragons swam in nearshore waters off the coast of Zealandia when it was located in the high latitudes of about 66°S. Land was not far distant - the sediments are rich in terrestrial silts, charcoal, pollen and bones of dinosaurs. The local vegetation was evergreen temperate rainforest with Huon Pine, other podocarp conifers, southern beech and common tree ferns, indicating a mild temperate climate lacking severe frosts.

REFERENCES

174, 400, 423, 424, 425

Waipara River, New Zealand

THIS PAINTING SHOWS A SCENE IN THE LATEST Cretaceous seas that once covered part of New Zealand. Here, well known from the South Island's Waipara River deposits, is the large mosasaur *Prognathodon waiparaensis*, which was described by S.P. Welles and D.R. Gregg in 1971 from a superb specimen now in the Canterbury Museum. This giant marine reptile must have been a fearsome sight, with a body up to 11 m long. Its 1 m jaws were lined with large conical teeth. In addition to rows of teeth on its upper and lower jaws, two rows of teeth on the roof of its mouth, all curved backwards, would have prevented prey escaping and enabled the reptile to tear off large chunks of flesh and swallow them whole. As the dominant predator of the seas at this time, it almost certainly would have preyed on elasmosaurs, turtles and large fish. Protostegid turtles were somewhat like modern leatherback turtles and were the first to show the modern body plan of marine turtles. Known from North America and Australia, they were first recorded from New Zealand by Joan Wiffen in 1981 based on eight fragmentary bones from the Mangahouanga Stream, Hawke's

Bay (see p. 88). Little is known of their diet, but individuals of *Notochelone* in Australia had fragments of the mollusc *Inoceramos* in the stomach. Flying overhead is a pterosaur. Although not actually known from the Waipara River deposits, they were likely present because pterosaurs are recorded from the Mangahouanga Stream deposits a short distance away in Hawke's Bay. ●

AGE

Late Cretaceous, late Haumurian (Maastrichtian), about 73 million years ago.

LOCALITY

Waipara River, North Canterbury, New Zealand.

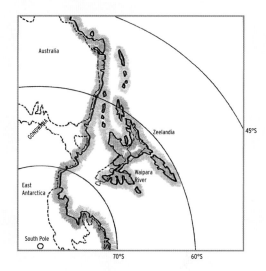

ENVIRONMENT

The late Cretaceous seas revealed by the Waipara River beds lay just offshore of Zealandia, which was a newly created land recently rifted from Gondwana. This site lay only about 100 km from the beds being deposited in Hawke's Bay just a few million years earlier.

REFERENCES

106, 173, 412, 426

Haumuri Bluff, New Zealand

CRETACEOUS

IN THIS SCENE, A LARGE TYLOSAURINE MOSASAUR, *Taniwhasaurus oweni*, attacks a much smaller pliosaur. Pliosaurs had a broad body with a short neck, a short tail and four large flippers. All pliosaur remains from Haumuri Bluff are fragmentary and, although they have been referred to five species since Sir Richard Owen described *Plesiosaurus australis* from this locality in 1861, in reality few are sufficiently complete to enable confident species-level identification. One of the described taxa is *Polycotylus tenuis*, named by Sir James Hector in 1874. If correctly placed in this genus, which is

otherwise known from North America, Russia and Australia, the Haumuri Bluff pliosaur would have had a long head at the end of its short neck. Its limbs were essentially paddles highly adapted for swimming and of little use for terrestrial locomotion. Hence, it is not surprising to find that pliosaurs almost certainly gave birth to live young, as did ichthyosaurs, which means these reptiles never had to leave the oceans.

Taniwhasaurus oweni was a large mosasaur, probably exceeding 7 m in length. Other members of the genus are known from Japan and James

Ross Island, on the Antarctic Peninsula. Its generic name derives from the 'taniwha', a supernatural aquatic creature in Māori mythology. With its huge head and many large teeth, it was undoubtedly the top predator in the late Cretaceous seas in this region of the world. Given its size, it could have killed even the large pliosaurs, which themselves were major predators capable of eating large and agile prey, probably including plesiosaurs. ●

AGE

Late Cretaceous (Lower-Middle Campanian), about 69 million years ago.

LOCALITY

Haumuri Bluff, Kaikoura, New Zealand. This was a tremendously important site for fossil collecting in the late 19th century and a key to developing the geological timescale in New Zealand.

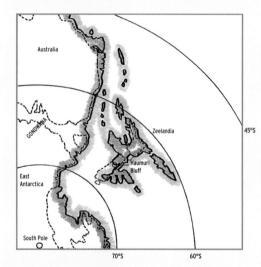

ENVIRONMENT

This Haumuri Bluff biota was deposited in a nearshore environment, on the Pacific margin of Gondwana towards the end of the Cretaceous period when New Zealand separated from other Gondwana fragments.

REFERENCES

70, 89, 169, 412

Tingamarra, Scene 1, Queensland
EOCENE

IN THE SHALLOWS OF THE TINGAMARRA SWAMP, two crocodilians lie in wait for incautious prey. Both are species of *Kambara*, the oldest known of the extinct Australasian mekosuchine crocodilians, and grew to lengths of around 2–3 m. Tingamarra's two early Eocene *Kambara* species share a generalised crocodilian body plan, not unlike that of the modern Saltwater Crocodile *Crocodylus porosus*. Their broad flattened skulls and large size suggest they were semi-aquatic ambush predators. Each of the two species had a distinctive dentition, probably reflecting different feeding strategies.

Kambara implexidens had an interlocking dentition while *K. murgonensis* had an overbite pattern. Because there is a predominance of adult- and hatchling-sized rather than intermediate-sized individuals among the fossils, the Tingamarra swamp may have been used as a nesting ground by one or both species. The relatively straight humerus and strong shoulder joint suggest that species of *Kambara* were better equipped for terrestrial locomotion (i.e. extended high walking where the body is held off the ground) and paraxial swimming (i.e. generating thrust by alternating strokes of the limbs) than are living crocodiles such as the modern Saltwater Crocodile. The Tingamarra waterbody was a low-energy freshwater lake, billabong or swamp. Intermittent bands of dolomite in the Tingamarra deposit suggest intervals of low water level that may have occasionally forced these crocodile species to come into direct contact with each other, an uncomfortable situation for most crocodilians.

Mekosuchine crocodilians were once endemic to Australia and islands in the South Pacific. They survived until the Pleistocene on mainland Australia and until the arrival of humans in Fiji, New Caledonia and Vanuatu. ●

AGE

Early Eocene, about 54.6 million years ago.

LOCALITY

Near Murgon, south-east Queensland.

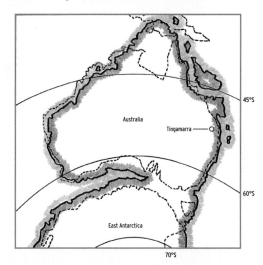

ENVIRONMENT

This was a time of warm wet conditions, during the globally recognised Early Eocene Climatic Optimum. Australia was still tenuously connected to an ice-free Antarctica, and it to South America, with broadleaf and southern conifer forests providing corridors for dispersal across these last vestiges of Gondwana.

REFERENCES

345, 369, 370, 433

Tingamarra, Scene 2, Queensland

EOCENE

IN TREETOPS ARCHING OVER A QUIET BILLABONG, the 80 cm madtsoiid snake *Alamitophis tingamarra* strikes at a small bird feeding its nestlings. In the process, it startles the tiny insectivorous marsupial *Djarthia murgonensis* foraging nearby. The now-extinct madtsoiids were constrictors found mostly in Australia, South America and Africa. Although species of *Alamitophis* were relatively small, some Australian madtsoiids the size of anacondas survived until the late Pleistocene – longer than anywhere else in the world. The mouse-sized marsupial *D. murgonensis* is the most primitive known member of the superorder Australidelphia, the group that includes all modern Australasian marsupials. This clade also includes the order Microbiotheria which today is represented by a single living species, *Dromiciops gliroides* from the Valdivian temperate rainforests of South America's southern Andes. Current wisdom suggests that the modern Australasian marsupial orders evolved from ancestors that dispersed from South America via Antarctica, sometime during the late Cretaceous to early Paleogene, and radiated prior to the late Oligocene. But the extremely plesiomorphic australidelphian morphology of *D. murgonensis* and the apparent absence of undoubted australidelphians from early Paleogene deposits in South America has been interpreted as evidence that they first evolved in Australasia, perhaps from a *Djarthia*-like ancestor. Hence, sometime after evolving in Australia from marsupials that dispersed from South America to Australia via Antarctica, at least one member of this Australian group must have migrated back to South America where they thrived and survive as *D. gliroides*. Tingamarra's unnamed passeriforms appear to represent a stem group that gave rise to the world's passerine songbirds. Today, passerines – which include among many other groups the wrens, lyrebirds, honeyeaters and crows – are the most diverse order of living birds. Although their phylogenetic and geographical roots remain contentious, given the fact that the Tingamarran passeriforms are the oldest known in the world, Australian origins for this group seems probable. ●

AGE
Early Eocene, about 54.6 million years ago.

LOCALITY
Near Murgon, south-east Queensland.

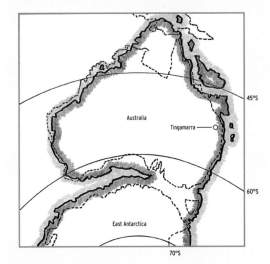

ENVIRONMENT
A forest fringing a low-energy freshwater lake, billabong or swamp. It was a period of warm wet conditions, during the Early Eocene Climatic Optimum. Australia was still tenuously connected to an ice-free Antarctica, and it to South America, with broadleaf and southern conifer forests providing corridors for two-way dispersals across these last vestiges of Gondwana.

REFERENCES
38, 58, 60, 148, 149, 347

Tingamarra, Scene 3, Queensland

EOCENE

AS DUSK FALLS, A 'GRACULAVID' SHOREBIRD wades in the shallows, chasing a last meal for the day. A bat *Australonycteris clarkae* makes its first insect catch for the night and a frog *Lechriodus casca* lets fly with its first call. A soft-shelled turtle *Murgonemys braithwaitei* surfaces for air. A marsupial *Thylacotinga bartholomaii* watches from the undergrowth. The 'graculavid' is one of two or three Tingamarra species representing a grade of primitive charadriiforms that is restricted to late Cretaceous to early Eocene deposits elsewhere in the world. The generalised postcranial morphology of these birds is characteristic of several groups of early neornithine birds in both the Northern and Southern Hemispheres. *Australonycteris clarkae* is one of the world's oldest bats. It is similar to other archaic early and middle Eocene bats from both the Northern and Southern Hemispheres and represents a grade of primitive stem-bats globally extinct by the late Eocene. It navigated using echolocation, like most bats do today, and probably hunted insects over the surface of the Tingamarra lake or in the fringing vegetation. The quoll-sized omnivore *T. bartholomaii* may belong to a family otherwise known from South America and/or Antarctica and suggests a trans-Antarctic migration route between South America and Australia extending well into the early Paleogene. It probably ate fruit, seeds, soft leaves and insects. *M. braithwaitei* belongs to a group of turtles called trionychids that have flexible (soft) shells, allowing them to deform as they squeeze into small spaces chasing prey such as crayfish, frogs and small fish. Trionychids died out in Australia in the Pleistocene, around 40 000 years ago, but relatives survive in New Guinea, Asia and Africa. *Lechriodus casca* indicates that at least some generic differentiation among Australian frogs predates the final fragmentation of Gondwana. Its only living Australian descendant, *L. fletcheri*, lives around pools or streams in rainforests and wet sclerophyll forests. ●

AGE

Early Eocene, about 54.6 million years ago.

LOCALITY

Near Murgon, south-east Queensland.

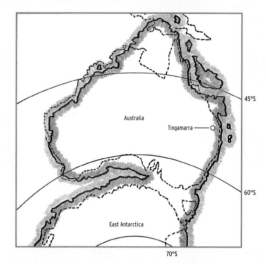

ENVIRONMENT

A forest fringing a low-energy freshwater lake, billabong or swamp. This was a time of warm wet conditions, during the Early Eocene Climatic Optimum. Australia was still tenuously connected to an ice-free Antarctica, and it to South America, with broadleaf and southern conifer forests providing corridors for dispersal across these last vestiges of Gondwana.

REFERENCES

20, 61, 163, 357, 399, 417

Duntroon, New Zealand

PETER SCHOUTEN 2011

SQUALODONTIDS, OR SHARK-TOOTHED dolphins, are large, long-snouted odontocetes (echolocating toothed whales) with typically high, triangular, often multicuspid and laterally compressed teeth. They were diverse in the late Oligocene to middle Miocene and were probably specialist predators on fish and, when they could catch them, birds. They survived until the early to middle Miocene when they appear to have been replaced by relatively more modern families of toothed whales. For many years squalodontids were thought to be a family from which modern toothed whales emerged. Now, however, several families are recognised within this group. True squalodontids (Squalodontidae) include species of *Squalodon* from the North Atlantic, *Phoberodon arctirostris* from the South Atlantic and several unnamed forms including some from New Zealand and possibly '*Prosqualodon*' *hamiltoni* from the early Miocene of Caversham, Dunedin. In the New Zealand region, most of the extinct shark-toothed cetaceans originally thought to be squalodontids are now regarded as members of Platanistoidea (e.g. *Waipatia maerewhenua*), a more diverse group that is the sister-clade of squalodontids. This broader group also contains the living platanistid Ganges River Dolphin *Platanista gangetica*. The platanistoid species illustrated here is the late Oligocene *Otekaikea marplesi* recovered from the Otekaike Limestone Formation in the Waitaki Valley of the South Island of New Zealand. It is known only from one partial skeleton that was described and named in 2014, and is one of only two species in the Waipatiidae, an extinct family of cetaceans unique to New Zealand and Australia. ●

AGE

Late Oligocene, 25.2 million years ago.

LOCALITY

Otekaike Limestone, Waitaki Valley, South Island, New Zealand.

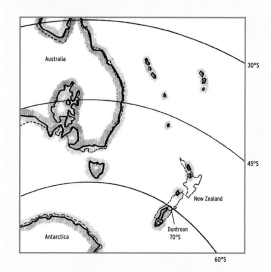

ENVIRONMENT

The environment where this cetacean and many others lived was a relatively shallow mid- to inner-shelf marine ecosystem.

REFERENCES

2, 95, 127, 376, 377, 378, 379

Jan Juc Formation, Victoria

OLIGOCENE

TWENTY-FIVE MILLION YEARS BEFORE TODAY'S ocean-addicted human surfers learned to ride the waves at Bell's Beach, Victoria, ancestral baleen whales and other fascinating marine creatures cruised these same waters. The timeless energetic waves still scour the tawny yellow cliffs at the back of the beach, gouging out of them the bones and teeth of those ancient ocean-loving creatures. *Janjucetus hunderi*, one of the largest whales that left skulls and postcranial bones on the sea floor of this region, has proved to be a most fascinating cetacean because it appears to represent a 'missing link' group between ancient toothed whales (odontocetes) and the baleen whales (mysticetes). The osteology of the skull reveals that it shared key derived features with all baleen whales. However, unlike later baleen whales, it had no baleen and instead retained well-developed teeth like toothed whales. It was about the size of a large modern dolphin but, unlike all toothed whales and like most if not all baleen whales, it evidently could not echolocate. Because it had very large orbits in the skull, it probably visually located its prey, which was likely to be fish, which it savaged much as killer whales do today. The group of smaller whales shown here foraging on the ocean floor are *Mammalodon colliveri*. Each was about 3 m long. They appear, on the basis of skull and dental morphology, to be another member of the 'missing link' group of pre-mysticete whales that contains *J. hunderi*. Both are now placed in the family Mammalodontidae, this name reflecting the fact that unlike all modern cetaceans they retained relatively conventional mammal-like dentitions with incisors, canines, premolars and molars. Other vertebrates shown here include the cusk eel *Ophidion granosum*, a weird group of eel-like fish that can burrow tail-first into the mud but spend most of their time hunting for food on the ocean floor. *Heterodontus cainozoicus* is a distant relative of the Port Jackson Shark in the family Heterodontidae and, like its modern counterpart, it may have been a nocturnal feeder on molluscs and echinoderms that were abundant on the ocean floor. *Megalops lissa*, flanking *J. hunderi*, was an ancestral tarpon (Megalopidae) that was an agile predator of fish in the open ocean. *Gadus refertus*, shown hunting near the bottom, is a predatory relative of cod fish (Gadidae) that would have eaten a wide range of foods near the bottom of this marine habitat. ●

AGE

Late Oligocene, about 25 million years ago.

LOCALITY

Coastal sandstone cliffs, Lorne-Queenscliff Coastal Reserve, Jan Juc, south-west of Torquay Surf Beach, Victoria.

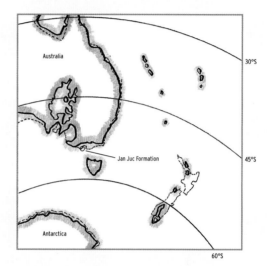

ENVIRONMENT

Nearshore seas off the coast of south-east Australia.

REFERENCES

44, 111, 112, 186, 223

Ditjimanka Local Fauna, Scene 1, South Australia
OLIGOCENE

CENTRAL AUSTRALIA'S EPHEMERAL FRESHWATER lake system in the Lake Eyre Basin was once home to *Australosuchus clarkae*, a 3–4 m crocodile that appears to have been a generalist aquatic predator. It probably ambushed prey that came to drink at the lake's edge. When *A. clarkae* lived, during the late Oligocene to early Miocene, these central Australian lakes may have been permanently filled with water supporting fish, turtles, platypus and even freshwater dolphins. The surrounding forests supported a diverse fauna including extinct sheep-sized ilariid marsupials, koalas, bandicoots, bats, possums of many kinds and flamingo-like paleolodids. This crocodile is known from several deposits in the Lake Eyre Basin that have produced skulls, dentaries and postcranial elements. The most common items are isolated teeth that were shed during life, and bony scutes that developed within the skin to act as defensive armour. Its teeth varied in size, shape and function along the tooth row, with some adapted for puncturing and holding prey and others with vertical ridges better suited for slicing and cutting flesh.

This crocodile was a member of the subfamily Mekosuchinae, a group endemic to the south-west Pacific region including the Australian mainland, Vanuatu, New Caledonia and Fiji. The earliest records of mekosuchines from the Australian early Eocene appear in the Tingamarra Local Fauna. Mekosuchines were common in Australia, as shown in mid to late Cenozoic fossil communities, until they became extinct sometime in the Pleistocene. However, until 3000–4000 years ago they survived on the islands of New Caledonia and Vanuatu. ●

AGE
Late Oligocene, about 25 million years ago.

LOCALITY
Palankarinna, Pinpa, Tarkarooloo and Ngapakaldi Lakes in the Lake Eyre Basin, South Australia.

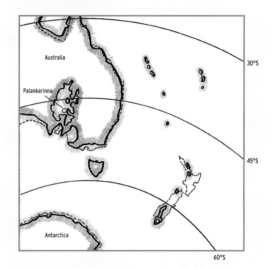

ENVIRONMENT
In the late Oligocene, in the Palankarinna and Pinpa lakes regions of the Lake Eyre Basin, permanent freshwater lakes were surrounded by humid scrubby rainforests dominated by *Nothofagus* species. These areas became warmer and much drier in the Neogene and Quaternary and are now arid/semi-arid deserts.

REFERENCES
429, 431

Ditjimanka Local Fauna, Scene 2, South Australia

OLIGOCENE

IN THE LATE OLIGOCENE, CENTRAL AUSTRALIA looked very different from today. The now seasonally dry Lake Eyre Basin was at that time dotted with freshwater lakes surrounded by forests dominated by *Nothofagus* (southern beech), casuarinaceans and myrtaceans. *Palaelodus pledgei*, shown flying above the lake, were tall slender birds with long legs and necks; they belonged to the extinct family Palaelodidae. These flamingo-like birds were in fact distantly related to flamingos. They were browsers that fed while swimming or standing in shallow water. In the forests lived archaic extinct marsupial ilariids (Ilariidae) and wynyardiids (Wynyardiidae), as well as the oldest known bandicoots (Peramelemorphia), koalas (Phascolarctidae) and burrowing bats (Mystacinidae). *Ilaria lawsoni* (far right) was a cow-sized quadrupedal marsupial whose selenodont teeth indicate it was a leaf-eater. Its postcranial remains suggest it was terrestrial with possibly some capacity for digging. Ilariids are known from the late Oligocene only and hence provide a useful marker for Australian fossil deposits of late Oligocene age. Relationships of this family to other diprotodontian marsupials are uncertain but they are at least distantly related to wombats (Vombatidae). *Muramura williamsi* (far left) was an extinct wynyardiid that was around the size of a dog and probably herbivorous. It is known from two articulated well-preserved skeletons. *Perikoala palankarinnica* (top centre), one of the oldest-known koala, is one of the extinct phascolarctids from the late Oligocene of central Australia. *Bulungu muirheadae* (lower far right), one of Australia's oldest-known bandicoots, is known from numerous dentaries and isolated teeth. This small insectivore or carnivore had a body mass of less than 250 g. A single tooth represents the burrowing bats (Mystacinidae) (lower right), a family that is today found only in New Zealand. All other members of this bat family appear to have been semi-terrestrial and it is likely that this central Australian species was too, feeding probably on both flying and terrestrial insects and other arthropods. ●

AGE
Late Oligocene, about 26 million years ago.

LOCALITY
SIAM and Tedford localities, Etadunna Formation, Lake Palankarinna, Tirari Desert, South Australia.

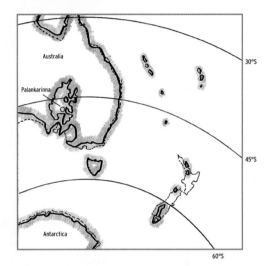

ENVIRONMENT
The late Oligocene scrubby rainforest forests that surrounded the waterbody that accumulated these fossils was dominated by casuarinaceans, myrtaceans and species of *Nothofagus* with araucariaceans and podocarpaceans. Sclerophyllous taxa including species of *Acacia* were also present at this time.

REFERENCES
12, 29, 151, 160, 165, 223, 276, 277, 382, 393, 437

Riversleigh, Scene 1, Queensland
OLIGOCENE

ALTHOUGH THE LATE OLIGOCENE FOREST
teemed with edible creatures, it was hard for
carnivores of any kind to resist trying to steal each
other's kill. Here the wombat-sized marsupial
lion *Wakaleo schouteni* (named in honour of the
famous palaeoartist who was unaware that he was
rendering his own Oligocene avatar) has decided
to challenge the fox-sized thylacinid *Nimbacinus
dicksoni* for access to the balbarine kangaroo that
it has just killed. Given that this species of *Wakaleo*
at about 23 kg was probably four times the
bodyweight of *N. dicksoni* and had more formidable
killing teeth, it would not have been much of a
contest. The thylacinid would have had to retreat
or risk becoming part of the marsupial lion's next
meal.

This scene also focuses on a controversy about
how some of the limestones in the Riversleigh
World Heritage Area actually formed. While some
clearly formed in pools in caves (replete with
speleothems such as flowstones and other cave
formations) and freshwater lakes (lacustrine
limestones), it has been argued that some of the
late Oligocene limestone may have formed in
slow-moving carbonate-rich stream water that
deposited limestone whenever it encountered
objects like pebbles, bones or even leaves to create
what are called tufa dams. As the dam walls grow,
they back up more water, and bones and other
objects are cemented into the accumulating layers
of limestone in these trapped pools. Well-studied
tufa dams of this kind occur today at points

along the Gregory River which flows through the Riversleigh World Heritage Area. This may have been how some of the late Oligocene limestones in Riversleigh originally formed, but more extensive study of these limestones is required before this hypothesis can be adequately tested.

Fortunately, whatever the mechanisms involved in their formation, many different kinds of limestone at Riversleigh were capable of preserving some of the most remarkable fossils ever found. In addition to bones and teeth, they include soft tissues of plants and animals, cells and, incredibly, even nuclei within those cells. ●

AGE
Late Oligocene, about 25 million years ago.

LOCALITY
Hiatus Site, D Site Plateau, Riversleigh World Heritage Area, north-west Queensland.

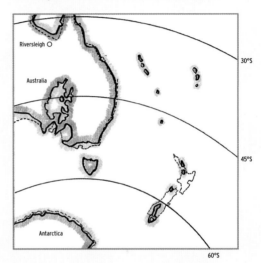

ENVIRONMENT
There is a lack of certainty about details of the plant communities in the Riversleigh area during the late Oligocene. However, overall structure of the fauna would suggest this lowland region had large freshwater lakes surrounded by forests. The latter may have had some areas with less dense cover and been cooler than the subsequent biodiverse rainforest communities that came to dominate the Riversleigh region during the early and middle Miocene. Topographically, there were islands of much older Cambrian marine limestones and Precambrian quartzites jutting up above the trees and lakes.

REFERENCES
25, 27, 141, 142, 144, 240, 266

Pinpa Local Fauna, South Australia
OLIGOCENE

LAKE PINPA IN NORTHERN SOUTH AUSTRALIA has been a fascinating source of many weird and wonderful late Oligocene beasts, not the least of which has been the somewhat wombat-like *Mukupirna nambensis*. In 1973, the late Dick Tedford from the American Museum of Natural History put a team together to follow-up his 1971 exploration for fossils at Lake Pinpa, east of the Flinders Ranges. It was fortunate timing, as sands that normally cover the dry salt lake's surface had been blown away, thus exposing the ancient clays that had accumulated 25 million years ago in a vast freshwater lake. Animals that lived in or on the waters of this lake included three kinds of stiff-tailed ducks (e.g. *Pinpanetta fromensis*, bottom left) and flamingos *Phoeniconotius eyrensis* (centre left). There were also freshwater dolphins. Stranger mammals that lived in the forests surrounding this lake, such as the 100+ kg *Mukupirna nambensis*, sometimes become mired in mud near the shore or drowned and settled into the gooey clay of the ancient lake bottom, where they were transformed into articulated but jumbled and often crushed collections of bones. Those who were on that expedition (which included this book's co-author Michael Archer) found that by gently pushing slender metal rods into the dry salt lake surface – a process that seemed a bit like acupuncturing the skin of mother Earth – they could sometimes detect hard objects below. These often proved to be the bones of skeletons that were not yet exposed at the dry lake's surface. One of these was the skeleton of *M. nambensis*. It was quarried out as a large mass, secured in wrappings of plaster of Paris and transported with many other casts collected that year at Lake Pinpa to the American Museum of Natural History. The skeleton was carefully recovered from the clay-filled cast and ready for study but, sadly, Dick Tedford died before he could describe it scientifically. Years later, in 2020, the specimen was finally published by a team of palaeontologists led by Robin Beck and Julien Louys. ●

AGE

Late Oligocene, about 25 million years ago.

LOCALITY

Namba Formation, Lake Pinpa, north-eastern central South Australia.

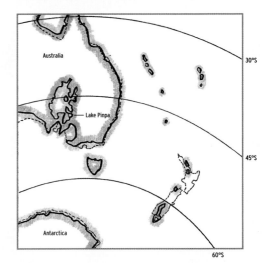

ENVIRONMENT

The ancient lake was enormous and must have once extended to the ocean to the south, enabling dolphins to reach this area and adapt to freshwater. The forest surrounding the lake was filled with ringtail possums, koalas and other arboreal animals as well as creatures that lived on the forest floor such as egerniine skinks. An extinct eagle is also known from this deposit. The forest itself was probably a type of scrubby rainforest.

REFERENCES

39, 51, 125, 239, 372, 388

Ericmas Local Fauna, South Australia

Freshwater dolphins and lungfish once inhabited an extensive lake system in central Australia that covered vast areas both east and west of the Flinders Ranges. Fossils recovered from late Oligocene sediments in the Lake Frome area of the Lake Eyre Basin include specimens of an extinct group of dolphins known as eurhinodelphids. These cetaceans are recorded globally from fossil deposits in the late Oligocene to the middle Miocene, but the central Australian ones are the first known to have occupied freshwater lake environments. They may be an indication that this gigantic inland lake had a direct but perhaps ephemeral connection southwards to the Southern Ocean. Cranial fragments including toothed rostra and ectotympanics as well as vertebrae reveal that these dolphins, which were about 3–4 m long, had very long snouts. Their small peg-like teeth indicate that, like many modern dolphins, they probably ate small fish. The long snout suggests that they used it to swipe at fish, like modern swordfish do. Other aquatic vertebrates in the same deposit include ray-finned fish, chelid turtles, mekosuchine crocodiles and at least four or five lungfish species, some of which reached more than 3 m. Lungfish, which can breathe air to supplement oxygen exchange via the gills, have been well represented in both marine and freshwater Australian fossil deposits since at least the Devonian. Also from this site, a single tooth represents the extinct toothed platypus *Obdurodon insignis*. ●

AGE

Late Oligocene, about 25 million years ago.

LOCALITY

Ericmas Quarry and South Prospect Quarry, Namba Formation, Lake Namba, Frome Embayment, Lake Eyre Basin, north-eastern central South Australia.

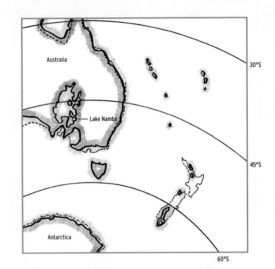

ENVIRONMENT

At this time, the Lake Frome area was inhabited by a range of aquatic vertebrates including ray-finned fish, lungfish, chelid turtles, mekosuchine crocodiles and dolphins. The habitat surrounding the lake would have been scrubby rainforest that contained a wide range of mammals, birds and other animals.

REFERENCES

23, 125, 203, 223

Riversleigh, Scene 2, Queensland
OLIGOCENE

A 4 M CROCODILE LUNGES FROM A FOREST LAKE where a marsupial *Ngapakaldia bonythoni* and rail *Australlus disneyi* drink at the water's edge. One of four croc species in this deposit, *Baru wickeni* was a member of the subfamily Mekosuchinae which was once common in Australia and some islands in the south-west Pacific – the last mekosuchine died out only 3000–4000 years ago in New Caledonia and Vanuatu. The 'cleaver-headed croc' *B. wickeni* had a very deep box-like skull and long blade-like recurved teeth that were probably used in a slicing cleaver-like fashion to kill its prey. It has been argued that *Baru* species were semi-aquatic, ambushing prey both on land and at the water's edge. This one was almost as large as the living Saltwater Crocodile *Crocodylus porosus* of northern Australia and south-east Asia, a crocodyline crocodile. It would have taken large prey including other reptiles, birds and mammals such as the cow-sized *Ngapakaldia bonythoni*. This primitive diprotodontoid, a distant relative of living wombats and the koala, was widespread during the late Oligocene. It lived on the shores of inland lakes from central Australia in the south to Riversleigh in the north, probably feeding on leafy vegetation on the forest floor and near the water's edge. The rail *A. disneyi* may have foraged in the same areas for small animal prey as well as shoots of reeds and other soft plants. This bird was a member of the family Rallidae, one that appears to be at least distantly related to the living, globally widespread swamphens (species of *Porphyrio*). ●

AGE

Riversleigh Faunal Zone A, late Oligocene, about 26-23 million years ago.

LOCALITY

White Hunter Site, D Site Plateau, Riversleigh World Heritage Area, north-west Queensland.

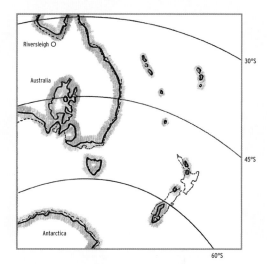

ENVIRONMENT

In the late Oligocene, the Riversleigh area was a limestone karst landscape covered by forest. Knowledge of the Riversleigh Faunal Zone A suggests a forest environment but one that may have been less dense than the subsequent rainforests that covered this area during the early and middle Miocene. Internally drained lakes were widespread throughout the area.

REFERENCES

46, 245, 368, 428, 448

Riversleigh, Scene 3, Queensland

BARAWERTORNIS TEDFORDI IS THE SMALLEST known member of the Dromornithidae, a family of large to gigantic flightless birds unique to Australia. It was about the size of a cassowary. Like the Northern Cassowary, which is a member of the unrelated family Casuariidae, its powerful legs would have been capable of moving it rapidly through the undergrowth when necessary, such as when threatened by predators. Such a predator may well have been *Ekaltadeta ima* (far left), a most unusual type of kangaroo belonging to the family Hypsiprymnodontidae, the only living

member of which is the Musky Rat-kangaroo *Hypsiprymnodon moschatus*, a small galloping (non-hopping) omnivore restricted to the tropical rainforests of north-east Queensland. *Ekaltadeta* had enormous, serrated premolars and powerful molars that indicate it specialised on animal flesh – it was a carnivorous kangaroo. *Ganawamaya couperi* (far right) was one of several very different kangaroos belonging to the archaic family Balbaridae. Like hypsiprymnodontids but unlike any modern kangaroos, it had a hallux (first toe) on the hind foot. Aspects of the hind leg suggest it would not have been as good a hopper as modern kangaroos. *Ganawamaya* were probably browsers as well as, perhaps, omnivores, a bit like modern tree kangaroos. *Onirocuscus silvicultrix*, a member of the possum family Phalangeridae, is an extinct relative of modern brushtail possums (species of *Trichosurus*) and cuscuses (species of *Phalanger* and several other genera) which live today in the forests of Australia, New Guinea and Sulawesi. *Onirocuscus* was about 2–3 kg (similar in size to a small modern brushtail possum), probably arboreal, nocturnal or at least crepuscular, and omnivorous, eating leaves and fruits but also small birds and reptiles. It would have had large eyes and a strong prehensile tail. ●

AGE

Riversleigh Faunal Zone A, late Oligocene, about 26-23 million years ago.

LOCALITY

White Hunter Site, D Site Plateau, Riversleigh World Heritage Area, north-west Queensland.

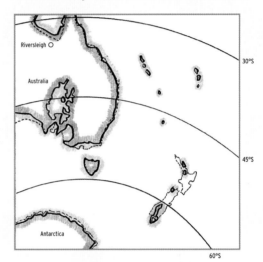

ENVIRONMENT

Many of Riversleigh's late Oligocene local faunas (Faunal Zone A) are characterised by the high proportion of aquatic and large-bodied animals compared to most of the younger Riversleigh faunal assemblages. The many large freshwater crocodiles, turtles, fish and frogs suggest that there were substantial bodies of water in the immediate area of the fossil deposits. There is no evidence for transportation of the fossils, which suggests they were relatively passively accumulated in local pools, lakes or caves with no involvement of river transport.

REFERENCES

18, 69, 81, 88, 188, 200, 279, 324

Riversleigh, Scene 4, Queensland

OLIGOCENE

BADJCINUS TURNBULLI WAS A CAT-SIZED thylacine, approximately 2.5 kg in weight, a carnivorous marsupial related to the recently (1936) extinct 'Tasmanian Tiger'. At least nine species of thylacines, all in the family Thylacinidae, are known from the Oligo-Miocene, ranging from kitten- to wolf-sized taxa. All were carnivores, with long snouts and cheek teeth characterised by oblique to longitudinal slicing blades. *Badjcinus turnbulli* is the oldest known member of this family. It is also one of the least specialised members and is close to the base of the thylacine tree. Its size suggests a diet of small vertebrates including mammals and birds such as the rail *Australlus disneyi*. Other non-passerine groundbirds from Riversleigh include the stork *Ciconia louisebolesae*. Storks are large birds that had a global distribution. This Riversleigh species was probably similar in size to a modern white stork and thus had a 2–5 kg body mass. It was probably carnivorous, eating frogs, fish, insects, earthworms, small birds and small mammals that lived around the ponds and lakes that were abundant at this time at Riversleigh. *Namilamadeta albivenator* was a dog-sized (20 kg) quadrupedal marsupial in the family Wynyardiidae, the last member of which went extinct about 15 million years ago. Wynyardiids were distantly related to wombats and koalas among living marsupial groups. This species is known from late Oligocene and early Miocene deposits of Riversleigh. Its dentition suggests that it was a herbivore, probably feeding on leaves from shrubs and trees. ●

AGE
Riversleigh Faunal Zone A, late Oligocene, about 26-23 million years ago.

LOCALITY
White Hunter Site, D Site Plateau, Riversleigh World Heritage Area, north-west Queensland.

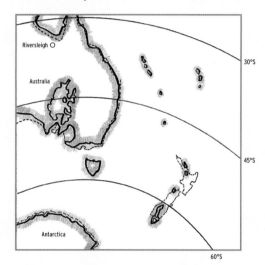

ENVIRONMENT
The White Hunter Local Fauna is considered to have lived in a forest environment. Although it has been suggested that the main limestone that contains most of the Faunal Zone A assemblages was formed in lakes trapped by tufa dams, there is some uncertainty about how these deposits formed. This particular deposit may well have been formed in a large cave.

REFERENCES
25, 62, 268, 300, 341

Wynyard Local Fauna, Tasmania
MIOCENE

THE EARLY MIOCENE DEPOSITS EXPOSED ALONG the shore at Wynyard, Tasmania, have produced millions of fossils including remnants of corals and the shells of marine molluscs, brachiopods and echinoderms. Vertebrate fossils have occasionally also made an appearance. One such specimen from a locality called Fossil Bluff is the holotype of *Prosqualodon davidis*. It was originally represented by a skull (now missing) and an almost complete skeleton. There are, fortunately, casts of many of the missing elements in the Tasmanian Museum in Hobart. This marine mammal was a toothed whale in the cetacean family Squalodontidae, members of which have been called 'shark-toothed dolphins' because of the triangular shark-like form of their teeth that contrast with the usual peg-like teeth in most modern dolphins. These teeth tended to jut outwards and would probably have been visible when the mouth was closed. This squalodontid would have been about 2.3 m long and a carnivore, probably specialised for feeding on other oceanic animals like fish, squid and birds. It may even, like killer whales, have been a predator of other cetaceans.

The Fossil Bluff Sandstone is also famous for producing the first known pre-Pleistocene mammal from Australia, the enigmatic *Wynyardia bassiana*, which must have been washed out to sea by a coastal river. Unfortunately, when the skull and partial skeleton eroded out of the sandstone, they slid down the cliff face teeth-down. As a result, when the specimen was found sometime around 1860 at the base of the cliff, it no longer had any of its teeth. Because palaeomammalogists have traditionally relied on teeth to help them work out evolutionary relationships, when the specimen was described in 1901 as a member of a new family of marsupials, the Wynyardiidae, its interfamilial relationships were unclear. However, based on comparisons of the skull and skeletal elements, it was then thought to probably be close to phalangerid possums. Because complete wynyardiid dentitions and skulls have been found at Riversleigh, for example, we now know that wynyardiids are in fact vombatoid marsupials most closely related among living animals to wombats and koalas. ●

AGE

Fossil Bluff Sandstone, early Miocene, about 23 million years ago.

LOCALITY

Fossil Bluff Site, near Wynyard, Tasmania.

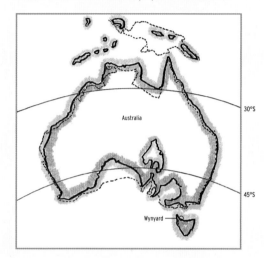

ENVIRONMENT

The seas just off the coast of south-eastern Australia would have been relatively warmer than they are today. Based on pollens found in the same formation, the habitat at the same time on the adjacent land would have been *Nothofagus*-gynmosperm evergreen rainforest.

REFERENCES

9, 124, 126, 223, 234

Kutjamarpu Local Fauna, South Australia

WHEN PAUL LAWSON WALKED ALONG THE eastern edge of Lake Ngapakaldi in 1962, as a member of Reuben Stirton's expedition in search of pre-Pleistocene fossils in central Australia, he spotted pieces of fossil turtle shell lying on the surface of the dry salt-encrusted mud. In the days that followed they discovered that these had been part of an ancient streambed that they named the Wipajiri Formation, and named the site the Leaf Locality because of a fossil leaf found in shales that overlay the stream deposit. When they excavated the stream deposit, lots of other teeth and bones were found including those of a small koala *Litokoala kutjamarpensis*, an archaic wombat *Rhizophascolonus crowcrofti* and the teeth of a seriously weird animal *Ektopodon serratus* (upper left), that they concluded may have been an archaic toothed monotreme. As subsequent research demonstrated, this was the first discovery of one of the most unconventional groups of short-faced possums, the Ektopodontidae. Its uniquely shaped molar teeth were formed by two rows of many tiny cusps that formed pseudo-lophs, a bit like the teeth of rodents but with the lophs oriented transversely rather than longitudinally. Despite subsequent discovery of teeth of many other kinds of ektopodontids in other fossil deposits, most aspects of these strange marsupials apart from their teeth and parts of the skull remain a complete mystery. They were always rare in the late Oligocene to Quaternary lowland rainforest communities in which they lived, until the last one vanished in the Pleistocene.

In 1971, palaeontologists Mike Woodburne, Bill Clemens, Mike Archer and Colin Campbell were excavating the Leaf Locality as part of another international team following up on Stirton's earlier investigations. Soon after opening the quarry, they noticed that Mike Woodburne was secretively working away at something he had found in the stream deposit. When he finally revealed his treasure, it turned out to be a spectacularly well-preserved lower jaw of one of the first leopard-sized marsupial lions. It was named *Wakaleo oldfieldi* (right). The generic name means 'little lion' while the species name honoured Brian Oldfield, the owner of Etadunna, the cattle station that included this region of the Tirari Desert. Did these marsupial lions hunt ektopodontids? No doubt if they found one like this foraging in the base of an epiphytic Birds Nest Fern (*Asplenium* sp.) for forest seeds or other things to eat, these carnivores would have been delighted to dine on such a rare but probably tasty rainforest inhabitant. ●

AGE

Early to middle Miocene, about 23–15 million years ago.

LOCALITY

The Leaf Locality, Wipajiri Formation (which contains a lower unit, the Basal Conglomerate, that formed in an ancient stream and an overlying clay, the Leaf Shale, that formed in a lake). This Formation is exposed on the eastern edge of Lake Ngapakaldi in the Tirari Desert of north central South Australia, between the Cooper and Warburton Rivers.

ENVIRONMENT

Many species of ringtail possums and a koala strongly suggest the presence of a forest. The immediately overlying clay beds (the Leaf Shale) contain tree leaves that have not yet been taxonomically studied. However, the fact that about 15 species from the Kutjamarpu Local Fauna also occur in the early and middle Miocene rainforest communities of Riversleigh, including *W. oldfieldi* and possibly also *E. serratus*, suggests that a vast species-rich lowland rainforest community stretched from at least what is now north-west Queensland to this area of central South Australia.

REFERENCES

16, 76, 297, 298, 302, 303, 327, 374, 436

Riversleigh, Scene 5, Queensland

EARLY MIOCENE

ONE OF THE MOST WIDELY DISTRIBUTED AND long-lived marsupial species was the large, perhaps sexually dimorphic diprotodontid *Neohelos tirarensis* (left). It is known from fossil deposits in Queensland, the Northern Territory and South Australia, and lived 25–15 million years ago. Body mass estimates based on partial skulls and mandibles range between 115 kg to 250 kg. It was a browser and probably fed on the leaves of shrubs and trees. *Dromornis* sp. (right) was an extinct flightless bird belonging to the family Dromornithidae that stood approximately 2.5 m tall and weighed up to 250 kg. These gigantic flightless birds appear to be most closely related to ducks and geese. The evolutionary relationships, enormous size and possibly omnivorous habits of some earned these dromornithid birds the nickname 'Demon Ducks of Doom'. Another species from this deposit with a colourful nickname is Fangaroo, an extinct species of balbarid kangaroo with the scientific name *Balbaroo fangaroo* (far right). Approximately the size of a modern Agile Wallaby, it was initially described on the basis of a partial skull in which the curved upper fanglike canine teeth are more than twice as long as the adjacent incisors. They may have used these canines for display or defence in the same way as long canines are used by herbivorous Asian mouse deer today. The extinct lyrebird from Riversleigh, *Menura tyawanoides* (middle right), is the earliest known member of its family, the Menuridae. Modern lyrebirds are among the best known and largest of Australia's songbirds, most notable for their remarkable ability to closely mimic natural and artificial sounds such as the calls of other birds or even the sound of a flushing toilet. ●

AGE

Riversleigh Faunal Zone B, early Miocene, about 20 million years ago.

LOCALITY

Upper Site, Godthelp Hill, D Site Plateau, Riversleigh World Heritage Area, north-west Queensland.

ENVIRONMENT

Upper Site is one of Riversleigh's richest fossil deposits. From a limestone deposit barely 2 m³ in size, at least 60 different fossil mammal species have been recovered. It appears to represent a cave deposit in which the remains of thousands of animals accumulated in a pool on the cave floor. This was a period of warm wet conditions globally, and at Riversleigh a species-rich closed canopy forest cloaked a limestone terrain riddled with caves.

REFERENCES

52, 59, 83, 274, 278, 373, 438

Riversleigh, Scene 6, Queensland

EARLY MIOCENE

ONE OF THE MOST ENIGMATIC PREHISTORIC
marsupials yet discovered is *Yalkaparidon coheni*,
a small (100–250 g) marsupial known only from
the late Oligocene to early Miocene forests of
Riversleigh. Its highly specialised dentition has
earned it the common name Thingodonta ('donta'
meaning tooth in ancient Greek). It shares features
with the other marsupial orders (e.g. elongate
lower incisors like kangaroos and wombats, molars
similar to marsupial moles, cranial morphology
like bandicoots) but most of these appear to be
either primitive or convergently evolved features.

For this reason, it and another species in the
same genus, *Yalkaparidon jonesi*, were placed
in a separate order, Yalkaparidontia, the only
Australian marsupial order to have gone extinct.
There are, however, possibilities that it could turn
out to be a member of an order of marsupials
in South America, the Paucituberculata, a group
that includes caenolestid marsupials and some
of their equally distinctive extinct relatives, some
of which have been found in Eocene deposits in
Antarctica. Although this remains controversial,
it emphasises how strange this bizarre group of

Riversleigh mammals obviously is. It was probably insectivorous, but exactly what it ate and how, remains a mystery. Its large ever-growing incisors but very reduced (zalambdodont) molars suggest food with a hard outer surface but soft insides, such as worms, caterpillars or eggs. It has been suggested that *Y. coheni* might have been a mammalian 'woodpecker' similar to the Striped Possum of northern Australia and New Guinea or the Aye-Aye of Madagascar, using its incisors to tear open weakened branches of trees in order to extract soft insect larvae. Predators of Thingodonta would have included the possum-sized marsupial lion *Lekaneleo roskellyae*. This was one of the least specialised of the marsupial lion family but was nevertheless equipped with the characteristic enlarged premolars that formed carnassial blades used in slicing through the flesh and bones of its vertebrate prey. These small marsupial lions may have been equally at home prowling among the treetops as well as on the forest floor. ●

AGE

Riversleigh Faunal Zone B, early Miocene, about 20 million years ago.

LOCALITY

Upper Site, Godthelp Hill, D Site Plateau, Riversleigh World Heritage Area, north-west Queensland.

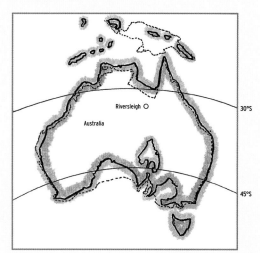

ENVIRONMENT

In the early Miocene, the Riversleigh area was covered in warm wet rainforest that cloaked the region's limestone and quartzite outcrops. There were no grasslands in Australia at this time. Caves that developed in the older limestone trapped unwary animals as they moved through the undergrowth. Pools in many of these caves enabled fine preservation of the most delicate bony structures as well as the uncrushed bodies of many different kinds of arthropods.

REFERENCES

14, 21, 37, 140, 143

Riversleigh, Scene 7, Queensland

EARLY MIOCENE

***NIMBACINUS DICKSONI* WAS A SMALL FOX-SIZED** thylacine (50 cm head–tail length; upper left), a carnivorous marsupial related to the 'Tasmanian Tiger' *Thylacinus cynocephalus*. It occurs in many fossil deposits at Riversleigh and is also known from Bullock Creek in the Northern Territory. A nearly complete skeleton of this species is known from AL90 Site on the Gag Plateau at Riversleigh. It was intermediate in size and carnivorous dental specialisations between species of *Badjcinus* and *Thylacinus*. A study of bite force (mechanical performance) revealed that this small thylacinid probably occupied a broadly similar ecological niche as that of the extant Quoll *Dasyurus maculatus*. On the basis of bite force and dental morphology, it probably hunted relatively large vertebrate prey that may have exceeded its own body mass. *Emuarius gidju* (right) was a large flightless bird that lived during the late Oligocene and early Miocene. It was first known from the Leaf Locality in central Australia and is common in Riversleigh's Faunal Zones A to C. It was closely related to emus and cassowaries. Its cassowary-like skull and femur and its Emu-like lower leg and foot suggested the nickname Emuwary or Cassemu, although it is technically closer to emus and may well be an ancestral emu. It probably browsed on shrubs and grasses as well as the occasional insect and small lizard. It was shorter than the modern Emu, standing about 1.5–1.8 m tall and weighing up to about 50 kg. It had smaller eyes than the modern Emu and was not as well adapted for fast running, which reflects the fact that it lived in more densely vegetated habitats. *Bellatorias* sp. cf. *B. frerei* (left) is similar to the extant Major Skink *B. frerei*, one of Australia's largest skinks. An agile climber, the living species is often found around rocky outcrops in well-vegetated areas including vine scrubs, rainforests and woodlands. Like the living species, it probably ate mainly insects but also some fruit. *Namilamadeta superior* (bottom) is known from a near complete skull from the early Miocene Upper Site. Sheep-sized, it was slightly larger than *Namilamadeta albivenator* from Riversleigh's White Hunter Site. It belongs to the family Wynyardiidae, the last member of which went extinct about 15 million years ago. Wynyardiids may have been ancestral to the first diprotodontoids, all of which went extinct during or just following the late Pleistocene. ●

AGE

Riversleigh Faunal Zone B, early Miocene, about 20 million years ago.

LOCALITY

Upper Site, Godthelp Hill, D Site Plateau, Riversleigh World Heritage Area, north-west Queensland.

ENVIRONMENT

The ecosystem that contributed to the Upper Site deposit was biotically highly diverse lowland rainforest. Compared with the slightly less biotically diverse ecosystems of the late Oligocene, these early Miocene forests were very wet and perhaps 1–2°C warmer. They were far more diverse than any of the rainforest communities in Australia today. However, comparable diversity today does occur in the Amazon and the lowland rainforests of Borneo.

REFERENCES

27, 184, 266, 300, 387, 449

Riversleigh, Scene 8, Queensland
EARLY MIOCENE

AS DAY FADES, THE SHEEP-SIZED 70 KG marsupial *Silvabestius johnnilandi*, one of Australia's largest arboreal mammals and an archaic member of the family Diprotodontidae, climbs a tree for the night. Here it will escape predators such as thylacines and crocodiles, and leisurely search the tree for fruits and other soft plant products. Like other extinct members of the family Diprotodontidae such as *Diprotodon optatum*, the last and largest member of the group, *S. johnnilandi* may well have travelled in herds. But unlike the larger diprotodontids, this herbivore's herds commonly moved upside-down in the crowns of the rainforest trees, rather like marsupial sloths or sun bears. In this highly unusual lifestyle, it was apparently similar to another even better-known diptorodontid from the middle Miocene of Riversleigh, *Nimbadon lavarackorum* (p. 140). Also shown here foraging at dusk are several species of leaf-nosed bats. These are typically slow but very manoeuvrable fliers that catch insects by hovering and gleaning, hawking or aerial pursuit. Although some roosted in trees, most spent the day in caves. Some but not all of these ancient bats have recognisable living descendants. *Rhinonicteris tedfordi* (upper left), for example, is the oldest known ancestor of the living Orange Diamond-faced Bat of northern Australia. In contrast, the evolutionary relationships of *Xenorhinos halli* (bat shown in the foreground) remain obscure but its broad deep snout with large nasal cavities, peculiar head posture and extremely short palate suggest a unique echolocation call and foraging strategy. The larger *Riversleigha williamsi* had broad crushing teeth, tall crests on its thickened skull and only moderately inflated ear bones, suggesting it used a lower frequency call to locate prey such as heavily armoured beetles. ●

AGE

Riversleigh Faunal Zone B, early Miocene, about 17 million years ago.

LOCALITY

VIP and Bitesantennary Sites, Bitesantennary Valley, D Site Plateau, Riversleigh World Heritage Area, north-west Queensland.

ENVIRONMENT

The early Miocene palaeoenvironments in the Riversleigh area appear to have been closed rainforest. Like today, the Riversleigh limestone (karst) landscape was riddled with caves that provided lairs and shelters for predators and prey, day roosts for bats, and death traps for many other animals that dropped in unexpectedly.

REFERENCES

26, 47, 156, 157, 158, 159

St Bathans, Scene 1, New Zealand

ON THE SHORES OF THE ANCIENT LAKE

Manuherikia, a few inhabitants of a lost subtropical fauna feed in the shallow margins of this huge (5600 km²) freshwater body. In terms of both diversity and number of individuals, birds dominated this aquatic ecosystem, as they do today in many areas of New Zealand. Waterfowl were particularly diverse, with nine species. Here we see the as yet unnamed progenitor of New Zealand geese (species of *Cnemiornis*, a relative of Australia's Cape Barren Goose, with three goslings. Swimming in the water is a pair of Stiff-tailed

Ducks *Manuherikia lacustrina*, a bird about the size of a Shoveler. Like other stiff-tailed ducks, this Miocene species dove for its food. Peering into the water from its perch on a log is a heron *Matuku otagoense*, a bird that was about the size of a White-faced Heron but with slightly shorter legs. It and a co-occurring small fossil bittern are both the oldest and only known herons from the fossil record of Australasia. With its webbed feet, long legs and long neck, the flamingo relative *Palaelodus aotearoa* searches for prey in the shallow waters. Prowling around the margins of the lake is a crocodilian up to 3 m long. It may be a mekosuchine crocodile, possibly related to species known from Fiji, New Caledonia and Australia. This New Zealand species occurred further south than any of the world's other crocodilians. ●

AGE

Late early Miocene, about 19-16 million years ago.

LOCALITY

Map shows the continent Zealandia, emergent land at the time, and some of the modern-day coastlines.

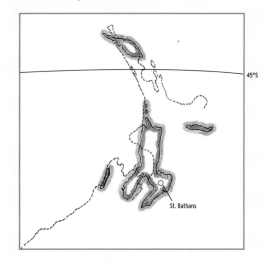

ENVIRONMENT

A subtropical habitat containing eucalypts, casuarinas, cycads and palms surrounded the lake's margin. Farther from the lake, taller forest with up to 60 species of laurel and abundant podocarp species clothed the nearby hills. Seasonal drought affected these habitats and associated fires were relatively common.

REFERENCES

351, 454, 455, 458, 459

St Bathans, Scene 2, New Zealand
MIOCENE

IN THE WOODLANDS BACK FROM THE EDGES OF
palaeo-Lake Manuherikia, strange yet vaguely
familiar inhabitants of the lost St Bathans Fauna
go about their business on the forest floor. Here
foraging in the litter of eucalypt leaves we find
small diplodactylid geckoes similar to those in
New Zealand today. Nearby is something very
much larger: an ancestral Tuatara. These are
sphenodontid reptiles, sister to all lizards. Today,
the last members of their order survive only in New
Zealand. Competing for invertebrates with these
reptiles were at least four species of burrowing

bats (in the family Mystacinidae) in two genera
(*Mystacina* [smaller bat] and *Vulcanops* [larger bat]).
Burrowing bats survive today only in New Zealand
where they spend almost as much time foraging
on the ground as in the air. Although perhaps the
most common bats in the St Bathans deposits,
they were not the only bats in this Miocene forest.
At least two other bats in two additional families
were present. But there is yet another mammal
in this picture (centre right, peeking out from a
crevice in the log). The size of a small rodent and
dubbed the 'Waddling Mouse', the relationships

of this (in fact these, because there are at least two kinds) tiny terrestrial mammal remain a mystery. Discovery of these enigmatic creatures came as a major surprise because, apart from introduced species, no non-flying terrestrial mammals had been known to have existed in New Zealand. Also scurrying about on the forest floor, but cloaked in feathers, is *Proapteryx micromeros*, a proto-kiwi. One-third the size of the smallest living Kiwi today and perhaps surprisingly able to fly a bit, this tiny early Miocene bird evolved into the living iconic national bird of New Zealand – the Kiwi. While this proto-kiwi or its ancestor must have flown in from afar, precisely where it came from is still uncertain. Like other Kiwi, *P. micromeros* was an omnivore and we see it here capturing a small leiopelmatid frog. This family of frogs is at least as old as the family of Tuatara and likewise restricted to New Zealand. ●

AGE
Late early Miocene, about 19-16 million years ago.

LOCALITY
Early Miocene deposits along the banks of the Manuherikia River, St Bathans, Central Otago, New Zealand.

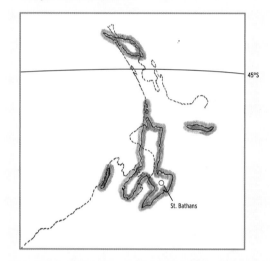

ENVIRONMENT
This seasonally dry, fire-affected woodland had much in common with those found in northern Australia today, with eucalypts, casuarinas and palms growing on the lowlands surrounding this huge lake. A large, braided river with white quartz gravel beds formed a delta where it entered the huge 5600 km² Lake Manuherikia.

REFERENCES
162, 164, 165, 195, 210, 457, 461

St Bathans, Scene 3, New Zealand

THE PALAEO-LAKE MANUHERIKIA ATTRACTED A vast array of different species from a range of different environments in and away from the water. Low hills and floodplains around the lake supported floristically diverse forests. Parrots had diversified into at least five different species, each of which exploited aspects of the rich and varied food resources available. But these woodlands existed in a seasonally dry climate; hence, especially in the drier months, many of the animals living in the surrounding habitats had to

come to the lake to drink. In the case of parrots, this was probably a daily occurrence. Here drinking on the margin of a small pool is a pair of a species of *Nelepsittacus* – small parrots that were one of at least four species of the nestorine radiation, the group which includes the familiar Kea and Kaka of today. Looking down on the smaller parrots is a strangely familiar, much larger, bird. Looking a bit like the modern Kakapo *Strigops habroptila*, we see the enormous *Heracles inexpectatus*. It was probably a member of the lineage that led the evolution of the Kākāpō. But this giant parrot, which has earned the nickname of 'Squawkzilla', was much larger than a Kākāpō. At about 80 cm tall and perhaps up to 7 kg in weight, it is the largest parrot that has ever been found anywhere in the world. ●

AGE
Late early Miocene, about 19-16 million years ago.

LOCALITY
Deposits along the banks of the Manuherikia River, St Bathans, Central Otago, New Zealand.

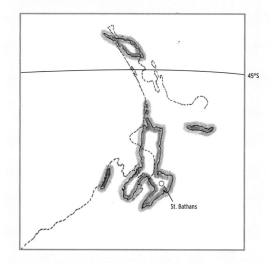

St. Bathans

45°S

ENVIRONMENT
The seasonally dry, floristically diverse woodland and forest environments around palaeo-Lake Manuherikia provided a rich food supply, from the fleshy fruits of podocarps to the nuts, cones and flowers of eucalypts, casuarinas, palms and, for carnivores, millions of incautious parrots.

REFERENCES
450, 460

Riversleigh, Scene 9, Queensland
MIDDLE MIOCENE

RIVERSLEIGH'S *OBDURODON DICKSONI* WAS A large platypus up to 60 cm long (head to tail). Older smaller species of *Obdurodon* are known from central Australia, and a closely related species, *Monotrematum sudamericanum* from the Paleocene of Argentina, is evidence that platypuses were once more widespread in Gondwana and not, as they are today, unique to Australia. Unlike the living Platypus, these fossil platypuses had functional cheekteeth. The well-developed rooted teeth of *O. dicksoni* suggest a more varied diet than that of the living Platypus,

perhaps including frogs and fish as well as insect larvae, yabbies and other crustaceans. It probably made burrows in the banks of rivers and streams where it laid its eggs. *Trilophosuchus rackhami* was a small mekosuchine crocodile perhaps 1.5 m long, from the early Miocene of northern Australia. It had a short deep head, large eyes and three longitudinal ridges along its skull (hence its name). It may have been terrestrial rather than aquatic and possibly even semi-arboreal, providing evidence that Australia did in fact have 'drop crocs'. Its neck musculature suggests it held its head above its body, as do varanid lizards. It probably would have preyed on small mammals, turtles, snakes and fish. The anatomy of the back of the skull suggests that feeding might have involved rapid side-to-side, up-and-down and rotational movement of the head in order to 'roll' prey in the way the Saltwater Crocodile does today. The turtle is *Pseudemydura*, a close relative if not the same species as the critically endangered Western Swamp Tortoise *P. umbrina* of Western Australia that lives in ephemeral swamps. The Riversleigh pygmy-possum *Burramys brutyi* is closely related to the living but critically endangered omnivorous Mountain Pygmy-possum *B. parvus* of the alpine zone in Victoria and New South Wales. The occurrence of this very similar ancestral form in a lowland rainforest environment has led to the suggestion that the living species, threatened by climate change, could be saved by translocating it into protected areas of lowland rainforest. A similar translocation strategy may be useful as a potential way to save the critically endangered Western Swamp Tortoise. ●

AGE

Riversleigh Faunal Zone C, middle Miocene, about 15 million years ago.

LOCALITY

Ringtail Site, Gag Plateau, Riversleigh World Heritage Area, north-west Queensland.

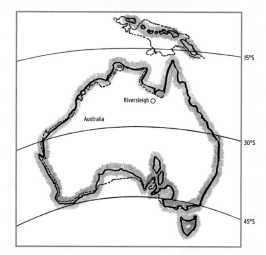

ENVIRONMENT

During the middle Miocene the Riversleigh area appears to have been a mosaic of lakes, pools and caves in a karst (limestone) terrain that was cloaked by closed species-rich gallery rainforest where species diversity was much higher than that of any rainforests of modern Australia.

REFERENCES

15, 25, 134, 275, 427, 430

Riversleigh, Scene 10, Queensland

MIDDLE MIOCENE

LITOKOALA KUTJAMARPENSIS **(PHASCOLARCTIDAE)** (top centre) was a small koala known from early/middle Miocene deposits in central Australia and middle Miocene deposits in Riversleigh. It was about half the size of a modern Koala *Phascolarctos cinereus*. Almost certainly arboreal, it probably fed on a variety of tree leaves. Specialisations in the middle ear indicate that it may have used loud vocalisations similar to those of the living species. *Onirocuscus reidi* (Phalangeridae; top left) is most closely related to species of *Trichosurus*. It was almost certainly omnivorous consuming fruits, leaves, birds and eggs whenever it had the chance. *Marlu kutjamarpensis* (middle left) was a ringtail possum

(Pseudocheiridae) found also in early Miocene deposits of central Australia. Its body mass has been estimated to be around 500 g. Its dentition suggests that it was a tree leaf-eater. Species of *Pildra* (bottom left) were much smaller ringtail possums (Pseudocheiridae) that may have fed on a wider range of plant foods, including lichens and mosses as well as soft leaves and blossoms. *Trichosurus dicksoni* (next to *M. kutjamarpensis*) was a brushtail possum (Phalangeridae). Like its modern rainforest counterpart, the Bobuck *T. caninus*, it was most likely an opportunistic omnivore that weighed about 1–2 kg. Species of *Cercartetus* (Burramyidae; lower right), known as pygmy-possums, are found today in a wide variety of habitats in Australia, New Guinea and Indonesia. As opportunistic omnivores they eat insects, flowers, fruit, nectar and pollen. *Macroderma godthelpi* (lower right), the carnivorous ghost bat, is related to the living Ghost Bat *M. gigas* of northern Australia. It would have used its good vision as well as echolocation to locate its prey, which would have been small mammals including other bats, small birds, lizards, snakes and insects.

Nimbadon lavarackorum is the large 70 kg tree-climbing herbivorous diprotodontid marsupial at the right. It is known from middle Miocene deposits at Riversleigh and Bullock Creek, Northern Territory. Features of its skeleton, such as the strong forelimbs, large claws and highly mobile shoulder and elbow joints, are strikingly similar to that of the living Koala, suggesting it had a similar arboreal lifestyle. However, given its much larger size, it may also have been capable of hanging upside-down when moving through or feeding in the canopies of the forest the way sloths or sun bears can do today. In terms of diet, it may have been primarily a fruit-eater but possibly also fed on soft leaves or flowers. As an arboreal fruit-eater, it probably helped to disperse the seeds of the rainforest trees. It is best known from many complete skulls and mandibles and at least two complete skeletons from the AL90 Site, a cave that acted like a natural pitfall trap, catching animals that fell out of the tree canopy overhead. The known specimens of this species represent individuals ranging in age from tiny pouch young to elderly adults. This mix of individuals of all ages suggests that they moved through the treetops in family groups or mobs, much the way kangaroos do today in open country. Compared with some of the better-known members of this family such as the gigantic *Diprotodon optatum*, it was on the smaller end of the size range of diprotodontids. However, along with *Silvabestius johnnilandi* (p. 130), these were the largest marsupials that have ever lived in the forest canopies of Australia. ●

AGE

Riversleigh Faunal Zone C, middle Miocene, about 15 million years old.

LOCALITY

Gag Site, Gag Plateau, Riversleigh World Heritage Area, north-west Queensland.

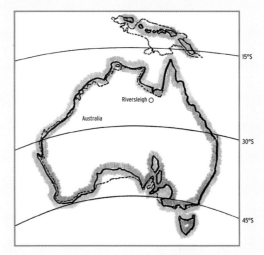

ENVIRONMENT

The Riversleigh area during the middle Miocene would have been a mosaic of lakes, pools and caves in a karst (limestone) environment. Palaeoecological studies suggest that the environment was dominated by species-rich rainforest.

REFERENCES

22, 48, 49, 50, 53, 54, 64, 88, 118, 155, 161, 230, 231, 278, 301, 339

Bullock Creek, Northern Territory
MIOCENE

CROCODILES ARE A COMMON DENOMINATOR OF most of Australia's Cenozoic palaeocommunities, and often were far more diverse in the past then they are at present. In the middle Miocene assemblages from Bullock Creek, one of the strangest forms was the crocodylid *Harpacochasma camfieldensis* which had an elongate, very toothsome mouth that made it look a bit like a living and presumably (there are controversies) unrelated Malayan False Gharial *Tomistoma schlegelii*. The diet of the False Gharial appears to include everything from fish to monkeys and even people, hence *H. camfieldensis*, while it looks like a fish-eater, may well have sampled more of the Bullock Creek biota than occurred only in the water. However, it would have been hard to pass up dining on the occasional very fleshy lungfish that were diverse in the middle Miocene waters of this area. One of these, *Neoceratodus* sp., may have been closely related to the living Queensland Lungfish *N. forsteri*. The Bullock Creek species of *Neoceratodus* coexisted with at least three other kinds of lungfish. The diversity of lungfish here suggests that the waters of this middle Miocene pool were never cold and were possibly subject to seasonal drying. Lungfish can survive shrinking deoxygenated pools by taking air into their lungs as well as by extracting oxygen from the water via their gills like other fish. In the adjacent forests, there was a wide range of animals similar to those that existed in the Riversleigh area at the same time. These included leopard-sized marsupial lions, horned turtles, fox-sized thylacines, archaic kangaroos and many individuals of *Neohelos stirtoni* which was a calf-sized diprotodontid marsupial. However, there appear to have been fewer arboreal mammals in the Bullock Creek faunal assemblage than there were in the middle Miocene assemblages of Riversleigh. ●

AGE

Middle Miocene, about 15 million years ago.

LOCALITY

Camfield Beds, Bullock Creek, Northern Territory. The limestone fossil deposits outcrop in a narrow belt about 50 km long near Victoria River in the Tanami Desert, Northern Territory.

ENVIRONMENT

The fossil assemblage suggests this region of the continent was beginning to become drier at the end of the middle Miocene. The thin enamel of all its mammalian herbivores suggests the vegetation was much less abrasive than it is in the same area today.

REFERENCES

52, 66, 204, 246, 273, 349, 429

Riversleigh, Scene 11, Queensland

***GANBULANYI DJADJINGULI* IS A QUOLL-SIZED** marsupial from the early late Miocene of Riversleigh (lower right). It shares some dental characteristics with the living Tasmanian Devil *Sarcophilus harrisii*; however, it may be a primitive dasyurid perhaps related to Riversleigh's *Barinya wangala* or belong to a separate carnivorous marsupial group. It is known only from a single upper heavily worn molar whose shape suggests this carnivore was a specialised bone-cracker. If so, it is the smallest bone-eating mammal known. *Yarala burchfieldi* (lower far right) is a tiny

(mouse-sized) bandicoot representative of a family thought to be near the base of the bandicoot radiation and structurally intermediate between bandicoots and dasyurids. It would have foraged in the forest leaf litter for insects and other small animals. This archaic bandicoot does not exhibit a close relationship to any of the other known types of bandicoots. *Palorchestes annulus* (upper right), another taxon known from a single molar, is the oldest known species of this genus. Over time, *Palorchestes* species (six are known) became larger and their teeth became more high-crowned and complex. Its youngest relative, the gigantic Pleistocene *P. azael*, had huge, compressed claws, powerful front limbs (adaptations to tearing bark or digging roots and tubers) and high-crowned teeth well suited to processing coarse or abrasive vegetation. To what extent features of this kind may have been present in *P. annulus* is unknown. Palorchestids are, in general, rare in the fossil record, suggesting that they were probably solitary animals. *Ganguroo robustiter* (upper left) is an archaic kangaroo that survived into the early late Miocene. Phylogenetic analyses using craniodental and postcranial characters suggest it belongs in the family Macropodidae, as the sister taxon to sthenurines and macropodines. *Tiliqua pusilla* (lower left) was a ground-dwelling insectivorous skink distantly related to the living Blue-tongued Lizard *Tiliqua scincoides*. It is the smallest known species of the genus. The Blue-tongued Skink has a large blue tongue that it uses as a bluff-warning to potential predators. ●

AGE

Riversleigh Faunal Zone D, early late Miocene, about 12 million years ago.

LOCALITY

Encore Site, Gag Plateau, Riversleigh World Heritage Area, north-west Queensland.

ENVIRONMENT

The Encore Local Fauna appears to represent a relatively more open forest palaeocommunity than those of the middle Miocene. Many of the animals in this faunal assemblage are larger than their relatives in the middle Miocene, a possible response to the beginning of drier conditions with less nutrient-rich vegetation. This is also the first time we see wombats *Warendja encorensis* that had evolved permanently growing teeth, probably an adaptation to dryness and the consequentially increasing sclerophylly of the vegetation.

REFERENCES

45, 80, 84, 267, 354, 387, 394, 462

Alcoota, Scene 1, Northern Territory
MIOCENE

THE ALCOOTA LOCAL FAUNA PROVIDES THE BEST opportunity to understand what Australia's land animal communities looked like in the late Miocene. While the fossil deposit appears to be missing many of the smaller creatures that must have been around at that time, bones of the larger ones occur in many thousands. One of the most common species represented is *Plaisiodon centralis* (lower left), a large (about 400 kg) diprotodontid marsupial that may have browsed on the leaves of shrubs while protecting its young from the predatory Powerful Thylacine *Thylacinus potens*.

A rarer and much larger (about 700 kg) diprotodontid was the browsing *Pyramios alcootense* (centre right). Its rarity may mean that it normally lived further away, resulting in fewer individuals ending up in the deposit at Alcoota. An even rarer but smaller (about 400 kg) diprotodontid was *Alkwertatherium webborum* (lower right). In some respects, this smaller diprotodontid might have looked a bit like a giant wombat. Perhaps both of these relatively rarer species were drawn to this location during a drought due to the large body of shallow water here, where the bones of trapped animals accumulated. *Ilbandornis lawsoni* (middle left) was one of several huge flightless, probably herbivorous birds in the family Dromornithidae that were common in the area and may have competed with diprotodontids for food. An even more awesome dromornithid known from Alcoota was *Dromornis stirtoni*. At more than 3 m tall and 500 kg in weight, this herbivore was the largest bird that ever existed. *Hadronomas puckridgi* (centre), at 1.5–2 m tall, was the largest kangaroo of its day. It may have been one of the earliest known short-faced kangaroos in the subfamily Sthenurinae, a group that went extinct in the late Pleistocene. Waiting in the water for any unwary bird or mammal was a giant mekosuchine crocodile (possibly an as yet unnamed species of *Paludirex*). If they got tired of waiting, crocs of this kind may well have ventured out of the water to chase down prey on the land. Very rare smaller animals have been found in this deposit, including a new kind of ringtail possum possibly related to the living Rock-haunting Ringtail Possum *Petropseudes dahli* of northern Australia. ●

AGE

Late Miocene, about 8 million years ago.

LOCALITY

Waite Formation, Alcoota, south central Northern Territory.

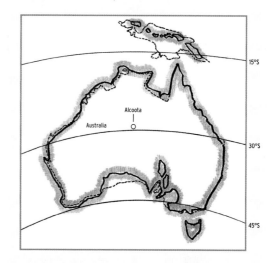

ENVIRONMENT

While today Alcoota is a vast area of grasslands and woodlands, in the late Miocene there would have been very little grass. However, there would have been plenty of shrubs and sparse woodlands to suit the culinary interests of the diverse large bird and mammal browsers. This was a time of drying and cooling all over the world. New Guinea, Australia's 'bumper zone', was rising as the Australian Plate crushed up against and dove under the Asian Plate, creating a significant rain shadow across Australia.

REFERENCES

51, 273, 324, 435, 465

Alcoota, Scene 2, Northern Territory

MIOCENE

The Alcoota Local Fauna includes one of the largest carnivorous thylacinids known, the Powerful Thylacine *Thylacinus potens*. At 1.5 m in length, it was about the size and build of a modern Grey Wolf *Canis lupus*. Any of the herbivorous mammals of its day would have been wary of this predator. At Alcoota, where it was rare, or at least rarely fossilised, it would have had a bit of competition from a dog-sized marsupial lion *Wakaleo alcootaensis*. It is possible that these marsupial lions hunted more in the denser bush or even in trees, leaving open-country predation to

the Powerful Thylacine. It is not clear how suited these thylacines would have been to long-distance running, and it is more likely that they were short-distance sprinters. Rather than running down animals, they probably pounced on startled prey that failed to see them hiding in the vegetation. We can be reasonably certain that there were other kinds of carnivorous mammals at Alcoota, but for some mysterious reason local carnivores rarely ended up in this deposit. But rarity is definitely not a feature of the Alcoota fossil record for *Kolopsis torus*. This bull-like diprotodontid marsupial would have stood nearly 1 m at the shoulder and been perhaps 1.5 m long. Although distantly related to wombats (which are also curiously missing from Alcoota given that they are well known from early, middle and late Miocene deposits at Riversleigh), its teeth indicate that it had a diet of leaves that were softer than the abrasive grasses which modern wombats and kangaroos consume. Given the abundance of this diprotodontid, it is possible that it moved through this area in large herds or mobs, a strategy that would have helped to protect the more vulnerable juveniles from becoming lunch for the Powerful Thylacine. ●

AGE

Late Miocene, about 8 million years ago.

LOCALITY

Waite Formation, Alcoota, south central Northern Territory. The enormously rich Alcoota fossil deposit extends laterally for 170 m. Bones and teeth are so abundant and often tightly packed in extensive bone beds that it can be difficult to excavate one fossil without breaking others that surround and overlap it.

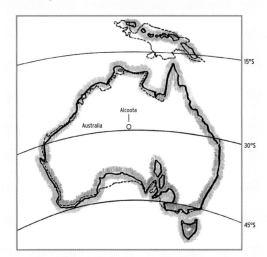

ENVIRONMENT

The Alcoota faunal assemblage includes the greatest variety of species of diprotodontids that has ever been described, which suggests that the vegetation of the area must have been equally diverse. It is probable that during the late Miocene this area was dominantly shrublands and woodlands. Although some grasses may have been present, there is no evidence for grasslands as such in Australia until the middle Pliocene.

REFERENCES

13, 24, 272, 273, 465

Beaumaris, Victoria

MODERN WHALES FALL INTO TWO CATEGORIES: toothed whales (e.g. the carnivorous porpoises and sperm whales) and baleen whales (e.g. the plankton-feeding right whales and humpbacks). Both groups have a long fossil history in Australia as they do in much of the rest of the world. The late Miocene *Balaena* sp. (shown here) from marine deposits at Beaumaris, Victoria, is one of the right whales (family Balaenidae). Although so far this baleen whale is only known from petrosals (bones that enclose the inner ear), these are very distinctive for the group. The non-flying fish-

eating penguins have an equally long fossil record, which is particularly rich in the marine deposits of the Southern Hemisphere. In Australia, their record extends from the Eocene to the present day. *Pseudaptenodytes macraei* (bottom left) is a large penguin in the family Spheniscidae represented by uniquely shaped upper arm bones (humeri) from Beaumaris. It was first described in 1970 by George Gaylord Simpson, a committed palaeomammalogist who regarded penguins as 'honorary mammals'. Among the strangest of the extinct birds from Australia, as elsewhere in the world, are the carnivorous pseudotoothed pelagornithids, such as the late Miocene *Pelagornis* sp. (shown flying and floating) from Beaumaris. Pelagornithids ranged in size from an albatross at the small end to giants with wingspans of 6 m. They probably used dynamic soaring to keep themselves airborne while sweeping over the sea in search of fish. Their strangest features were rows of bony projections along their jaws that looked and functioned like, but were not actually, teeth. The Beaumaris marine fossil deposit has also produced fossil whales, seals, fish and even rare land mammals. One of these is the medium-sized diprotodontid *Zygomaturus gilli*, which is similar to but a bit more derived than others from the Alcoota fossil deposit in the Northern Territory. The known age of the Beaumaris deposit helped to estimate the late Miocene age of the otherwise undated Alcoota deposit. ●

AGE
Late Miocene, about 6 million years ago.

LOCALITY
Black Rock Sandstone Formation, Beaumaris, Victoria.

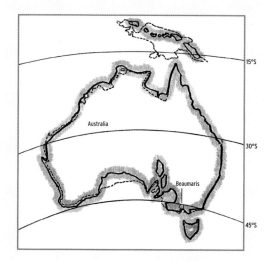

ENVIRONMENT
The Beaumaris fossil assemblage accumulated in a shallow invertebrate- and vertebrate-rich sea that extended across the southern part of Victoria. Habitats on the adjacent land area probably included relatively dry sclerophyll forests and woodlands.

REFERENCES
114, 124, 244, 286, 358, 385

Awe, Papua New Guinea
PLIOCENE

SOME OF THE BIGGEST MYSTERIES OF THE history of vertebrates of the Australasian region focus on New Guinea – when did it emerge from the sea, when was it colonised by land animals from Australia and did any that evolved in New Guinea spread back into Australia? The mid Pliocene Awe Local Fauna from the Otibanda Formation of Papua is the oldest known mammal-bearing fauna from the whole island. Three of the largest vertebrates in this fauna are marsupials although these tend, on average, to be smaller than closely related forms in Australia. The largest were the diprotodontids, an extinct group that may well have had its highest diversity in the Pliocene. Some have argued that the New Guinean forms are descendants of late Miocene invaders that would have resembled species from Alcoota. Others think they were a local endemic radiation that sprung from later more derived colonists. *Kolopsis rotundus* (upper left) is similar in dental form to *K. torus* from the late Miocene Alcoota Local Fauna, part of the reason for presuming that the source for these Pliocene Awe diprotodontids may have been late Miocene forms from central

Australia. The slightly smaller *Kolopsoides cultridens* (lower right) has very elongate upper premolar teeth, making its ancestry more of a mystery, although overall it too may well have descended from late Miocene Alcoota-type diprotodontids, in this case from something like *Plaisiodon centralis*. *Protemnodon otibandus* (upper right), a relative of larger 'giant wallabies' in the Pliocene and Pleistocene of Australia, was the largest Pliocene kangaroo known from New Guinea. It too has close relatives in Pliocene and Pleistocene deposits in Australia but there are no 'advanced' kangaroos like this in the Alcoota Local Fauna. This could be seen as support for the suggestion that there were at least two 'invasions' from Australia into New Guinea with the first (bringing in the diprotodontids) occurring during the late Miocene, and the second (bringing in species of *Protemnodon*) post-dating the start of the Pliocene. ●

AGE

Mid Pliocene, about 4-3 million years ago.

LOCALITY

Otibanda Formation, Awe, Papua New Guinea.

ENVIRONMENT

The Awe Local Fauna is not well enough known to give an unambiguous impression of the palaeoenvironment. All of the vertebrates reported so far, which, in addition to the three noted here, include rodents, a small dasyurid and a thylacinid, could be woodland or even grassland species bordering a stream. None is undoubtedly arboreal although limb bones suggest a possible tree kangaroo was present. Given the geographic and altitudinal position, it would have been at least a seasonally very well-watered palaeoenvironment.

REFERENCES

115, 120, 271, 294, 295

Chinchilla, Queensland
PLIOCENE

LEIPOA GALLINACEA IS A GIANT (4–7 KG)
malleefowl that lived in at least eastern Australia
during the Pliocene and Pleistocene. Males of its
smaller living relative, the Australian Malleefowl
L. ocellata, dig depressions in the ground which
they then fill up and mound over with leaf litter
and sand. As the leaves rot, they give off heat
which incubates the eggs. Although the males
manipulate the pile to regulate the temperature,
neither parent hangs around to see the young
dig their own way out of the pile and race off into
the bush – this is why juvenile mortality is high.

The mid Pliocene *Euryzygoma dunense* was a
bizarre knobble-headed large (about 2.5 m long)
herbivorous diprotodontid marsupial that was
probably the ancestor of the largest marsupial
known, the Pleistocene *Diprotodon optatum*. The
large bony processes that projected from the
cheek area may have supported cheek pouches
for holding food while it is being pulverised by
the teeth. Alternatively, because these processes
are much larger in some skulls than in others,
perhaps they served as sexual signals in males
because similar structures, such as horns and

antlers in ungulates, are often larger in males than females. The brain was surprisingly small, only about 8 cm long in a 60 cm long skull, with most of the cranial area comprising a mass of hollow sinuses leading to descriptors like 'air head' for these large herbivorous marsupials. They had the effect of enlarging the outer surface of the skull, which increased the areas of attachment for larger muscles of mastication. The short-faced kangaroo *Sthenurus notabilis* (centre) was one of several Pliocene sthenurine kangaroos, a group that became far more diverse in the Pleistocene before it vanished perhaps little more than 10 000 years ago. It is likely that this kangaroo was a browser, in contrast to more 'conventional' contemporary macropodine kangaroos such as the more diverse early species of *Macropus* which were grazers. By the mid Pliocene, grasslands were spreading for the first time across much of the continent. Leaping out of a tree towards the sthenurine is a Pliocene marsupial lion, *Thylacoleo crassidentatus*. It was slightly smaller than it's Pleistocene descendent, *T. carnifex*, but just as carnivorous. ●

AGE

Late Pliocene, about 4–3 million years ago.

LOCALITY

Main Gully System, Chinchilla Sands, Rifle Range, Chinchilla, south-east Queensland.

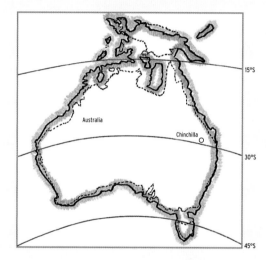

ENVIRONMENT

The vertebrate fauna from Chinchilla suggests a mixed environment that included grasslands (e.g. diverse grass-consuming kangaroos, wombats and rodents), trees (tree/shrub-leaf-eating kangaroos and koalas), shrubs (possibly the diet of some of the diprotodontids) and wetlands (turtles, crocs, fishes and wading birds). It was probably slightly better-watered than the area is today.

REFERENCES

10, 51, 63, 262, 318, 326

Kanunka, Scene 1, South Australia

PLEISTOCENE

WHILE TODAY'S SALTWATER CROCODILE *Crocodylus porosus* strikes fear in all northern Australian creatures, this 5 m mekosuchine crocodile, probably a species of *Paludirex* and possibly *P. vincenti*, may have been an even more dangerous predator if, like its smaller mekosuchine cousins, it could run down prey on land as well as leap up to grab them from the edge of the water. However, because it had an extremely broad snout, it is likely that this particular mekosuchine crocodile spent most of its time waiting in shallow water for thirsty diners to become dinner. Although

the oldest Australian published record for Saltwater Crocodiles is the mid Pliocene, from the Bluff Downs Local Fauna of north-east Queensland, crocodylines did not come to dominate Australia's northern waters until all of the mekosuchine crocs had, for reasons that are unclear, become extinct. At 3 m long, 2 m tall at the shoulder and perhaps 2000 kg in weight, the Hippopotamus-like browsing *Diprotodon optatum* is far and away the largest marsupial known. As currently understood, this giant lived only during the Pleistocene, dying out probably because of climate change perhaps

35 000 years ago. Early claims that it occurred in the Pliocene were based on a few aberrant teeth of *Euryzygoma dunense* from Chinchilla which do, however, indicate that *E. dunense* was probably the Pliocene ancestor of this Pleistocene giant. Fossil footprints of *D. optatum* punched into the mud of Lake Callabonna in South Australia preserve the eerie last steps of this geographically widespread giant as it sought water from drying pools on the lake's treacherous surface of sticky mud. In the background, the small pelican *Pelecanus cadimurka* (whose actual age, collected from nearby Cooper Creek, is uncertain) stands beyond the reach of the crocodiles, watching for fish stranded in the shallows. Many other waterbirds, including ducks and Australia's last flamingos, occurred in the lakes and river channels of central Australia during the Pleistocene. The geological record for pelicans in Australia extends back to the late Oligocene (25 million years ago) but a little earlier elsewhere, with 30 million years old species known from fossil deposits in France. The Pleistocene in Australia served up four major droughts alternating with milder wetter interglacial periods. Every time there was a new drought, each of which was more ferocious than previous ones, many species, particularly the megafaunal animals, vanished. By the time humans arrived on the continent, most of the megafauna had already disappeared. *Diprotodon optatum*, however, appears to have survived alongside humans for perhaps 25 000 years until at least 35 000 years ago. ●

AGE

Early Pleistocene, about 2.5 million years ago.

LOCALITY

Kanunka Local Fauna, Kutjitara Formation, Lake Kanunka, Tirari Desert, South Australia.

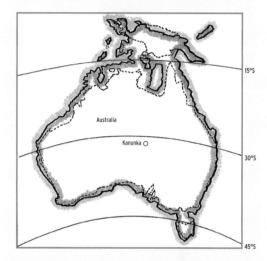

ENVIRONMENT

The river channels in this area of central Australia were probably running through country that was similar in terms of environmental conditions to those in the area today, with arid/semi-arid shrublands and grasslands. While Australia's oldest sand dunes are only about 1 million years old, widespread grasslands were commonplace after the mid Pliocene.

REFERENCES

253, 325, 344, 432

Kanunka, Scene 2, South Australia

PROCOPTODON GOLIAH, AT 2 M TALL AND 230 KG in weight, was the largest kangaroo known. Among its most interesting features was suppression of all but the huge fourth toe on the hind foot. In reducing the lateral toes, the species was becoming somewhat horse-like even though recent research suggests that this gigantic roo may have been a relatively poor hopper – if in fact it hopped at all. *Sthenurus stirlingi* (centre left), yet another of the diverse browsing sthenurine kangaroos all of which are extinct, at about 173 kg was one of the largest species of this genus but much smaller than *P. goliah*. Like all sthenurines, it probably had extended second and third fingers with elongate claws, useful for pulling down overhead leaf-laden branches. *Troposodon kentii* (shown bending down) was a macropodine kangaroo that belonged to a genus with controversial relationships within kangaroos in general. While originally noted to show striking similarities at the generic level to species of the living Banded Hare-wallaby *Lagostrophus fasciatus*, there were subsequent suggestions that species of *Troposodon* may represent a basal group within the subfamily Sthenurinae, which could mean that the Banded Hare-wallaby was in fact a surviving member of that subfamily. However, more recent research, while supporting the relationships between this hare-wallaby and species of *Troposodon*, does not support any special relationships to the sthenurines. But not all Pleistocene roos were giants; hare-wallabies like *Lagorchestes* sp. (bottom right), with a body length of perhaps 50 cm, was only about the size of a small dog. Like its recently extinct relative the agile Eastern Hare-wallaby *L. leporides*, it inhabited relatively dry country where it ate a range of ground plants. Flamingos (family Phoenicopteridae), while not part of the modern Australian biota, were once common around the shallow lakes of central Australia. Although the precise number of species and age ranges are uncertain, named Pleistocene species include the extant filter-feeding Greater Flamingo *Phoenicopterus ruber* (right foreground) as well as two extinct flamingos, *Xenorhynchopsis tibialis* (left foreground) and *X. minor* (background). ●

AGE

Early Pleistocene, about 2.5 million years ago.

LOCALITY

Kanunka Local Fauna, Kutjitara Formation, Lake Kanunka, Tirari Desert, South Australia.

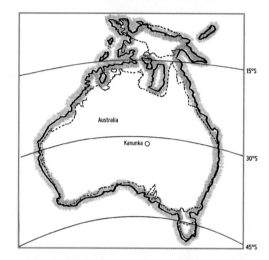

ENVIRONMENT

One of the most challenging aspects of the Pleistocene in central Australia was the unpredictability of conditions, ranging from sporadically well-watered woodland savannah during 'greenhouse' intervals to arid deserts during 'icehouse' intervals. Despite this, the Kanunka Local Fauna is surprisingly diverse, suggesting that at least some of the animals and plants of the early Pleistocene may have been more environmentally resilient than their surviving descendants. Nevertheless, most Kanunka species over 44 kg in weight went entirely or regionally extinct during the Pleistocene.

REFERENCES

71, 170, 187, 318, 445

Bunyip Cave, Victoria

A RUFOUS FANTAIL *RHIPIDURA RUFIFRONS* dances along the back of the Pleistocene Marsupial Tapir *Palorchestes azael* as this large herbivorous marsupial stoops to drink from a forest stream, while a modern Platypus *Ornithorhynchus anatinus* emerges from its bank-side burrow. The extinct palorchestids are often called marsupial tapirs because they have short retracted nasal bones in the skull, suggesting that they might have had a small mobile trunk like that of modern tapirs. *Palorchestes azael* was the largest of the palorchestids, with a body length of perhaps 2.5 m and some individuals weighing up to 1500 kg. Its teeth and skull morphology indicate that it may have been a browser, probably feeding on leaves, roots or tubers but also possibly the soft inner bark of trees, and perhaps even some invertebrates like insect larvae. Extensive areas on the lower jaw for the attachment of powerful jaw musculature suggest it might have had a long prehensile tongue, like those seen in living herbivores such as giraffes, for manipulating leaves and other vegetation. Powerful forelimbs and huge compressed claws may have been used to pull up shrubs, tear at the bark of trees or bring down branches to feed on foliage otherwise out of reach, while other aspects of its skeleton suggest it may have used a bipedal stance when feeding, perhaps like the ecologically similar extinct giant ground sloths of the Americas. The extraordinarily powerful forelimb was unable to be completely straightened and may have been held at about 100°. This limitation meant the animal would not have been able to reach higher than its head. Its fore and hind feet were almost certainly turned inwards, making this giant marsupial more or less pigeon-toed. None of the skeleton of the tail is known, so its length is a mystery. However, it is possible that the tail was short and solid, which would brace the animals while it was sitting on its haunches, freeing its powerful arms to do whatever they were able to do. ●

AGE

Pleistocene, about 1 million years ago.

LOCALITY

Bunyip Cave, Buchan Caves Reserve, eastern Victoria. One of the best-preserved specimens of this species was found there.

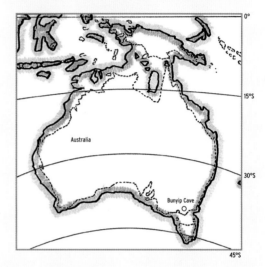

ENVIRONMENT

Open sclerophyll forests, probably with a grassy understorey, in many areas of eastern Australia.

REFERENCES

23, 117, 269, 270, 331

Eastern Darling Downs, Scene 1, Queensland

PLEISTOCENE

THE DIPROTODON *DIPROTODON OPTATUM* WAS the first fossil mammal named from Australia, and it remains one of the best known of the Australian megafauna. The first fossils were found in the Wellington Caves, New South Wales and described in 1838 by Sir Richard Owen of the British Museum. Diprotodon is globally famous for being the largest marsupial that ever lived. With an adult size of 3.8 m long, 1.7 m at the shoulder and some 2800 kg in weight, this was a rhinoceros-sized mammal with pillar-like limbs, a graviportal stance and inturned feet like those of wombats and palorchestids. It was widely distributed across mainland Australia, most commonly in the relatively drier inland parts of the continent rather than the wetter fringes. It is known only from the Pleistocene and serves as a marker species for deposits of that age. Its fossils include complete skulls and skeletons, as well as hair and foot impressions. There are trackways preserved at Lake Callabonna in central Australia and one skeleton that contains the remains of saltbush in its abdomen. This giant marsupial was probably a browser, feeding on shrubs, forbs and tree leaves in

Australia's open forests, woodlands and grasslands, eating daily as much as 100–150 kg of vegetation. Analyses suggest that this mega-herbivore migrated seasonally in herds or mobs in search of its preferred food, like many East African mammals do today. It co-existed with humans in Australia for at least 25 000 years. The timing and reasons for its extinction remain hotly debated. The two popular contenders are climate change or human activity, although there is no hard evidence that even a single individual was hunted by humans. *Diprotodon optatum* is the basis for the name of the family, Diprotodontidae. The Kookaburra *Dacelo novaeguineae* was originally given its species name on the mistaken assumption that it had been first seen in New Guinea, where it in fact does not exist. Its generic name *Dacelo* is an anagram of *Alcedo*, the Latin word for kingfisher. It is in fact a member of the kingfisher family Alcedinidae and they do include fishes in their diet when they can catch them. ●

AGE

Pleistocene, 2.6 million to about 35 000 years ago.

LOCALITY

The scene depicted is based on fossils from the eastern Darling Downs, south-east Queensland. *Diprotodon optatum* is known from many sites across Australia but not from New Guinea, Tasmania, Northern Territory or south-west Western Australia.

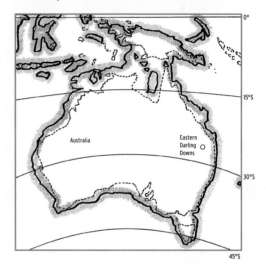

ENVIRONMENT

Semi-arid plains, savannahs and open woodlands. During the Pleistocene, climates in Australia oscillated between dry and more equable conditions. Sea levels were periodically lower than today at times when more of the world's water was tied up as ice at the poles and high latitudes.

REFERENCES

281, 311, 313, 463, 464

Eastern Darling Downs, Scene 2, Queensland

PLEISTOCENE

OWEN'S NINJA TURTLE *NINJEMYS OWENI* IS A large Pleistocene horned turtle from Queensland that had an estimated carapace (upper shell) length of 1 m and weighed around 200 kg. It resembled its larger Queensland relative, the Wyandotte Giant Horned Turtle, but the large pair of horns on its head extended to the sides rather than backwards as they tend to do in the Wyandotte species. Like other meiolaniid turtles, this meiolaniid turtle is thought to have been terrestrial and herbivorous. Meiolaniid turtles are also known from South America, where they are known from the Cretaceous to Palaeocene before they went extinct. The record in Australia isn't that old but it extends from the Eocene to the late Pleistocene and is more diverse than the South American record with species placed in several different genera. Precisely when and why the group became extinct on each of the Gondwanan continents they once occupied is not known, although they survived into the Holocene on some Australasian islands. *Ninjemys oweni* is shown here browsing alongside the sthenurine kangaroo *Procoptodon rapha*, one of many short-faced kangaroos that co-existed with longer-faced modern kangaroos during the Pliocene and Pleistocene. They specialised on a diet of leaves from trees and shrubs, with their robust skull and shortened face thought to be related to the need for large masseter and temporalis muscles necessary to browse on tough leaves and twigs. On each foot they had a single large toe or claw, somewhat similar in appearance to a horse's hoof. On each hand they had two very long fingers with large claws that may have been used to reach over their heads to pull branches down to the point where they could bite the leaves off. ●

AGE

Late Pleistocene, until at least 50 000 years ago.

LOCALITY

King's Creek, Darling Downs, Queensland.

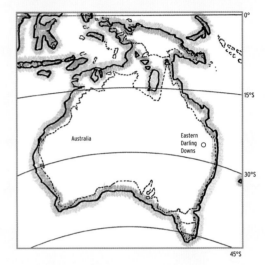

ENVIRONMENT

Streams and small rivers in this region were surrounded by sclerophyllous woodlands, scrublands and vine thickets. Grasslands may have been widespread but at a distance from the margins of the waterways.

REFERENCES

34, 131, 235, 305, 315, 318, 371, 439

Wellington Caves, Scene 1, New South Wales

***GENYORNIS NEWTONI* WAS THE LAST OF** Australia's large flightless dromornithids. Also known as mihirungs, thunder birds or even 'demon ducks of doom', these big birds are thought to be more closely related to ducks and geese than to ratites such as emus, cassowaries and ostriches. This particular dromornithid stood over 2 m tall and was very heavily built, with an estimated weight of 220–240 kg. It had small vestigial wings, and massive legs with hoof-like toes that indicate that this bird was well adapted for running. The beak was also massive and the lower jaw was unusually heavily ossified, indicating that considerable force could be used to break up its food. It appears to have been widely distributed in Australia during the Pleistocene, probably living mostly on the open plains and in open forests. Another heavyweight of this time was the bull-sized *Zygomaturus trilobus*, which weighed in at 500 kg or more, was 2.5 m long and stood about 1.5 m tall. This diprotodontid had prominent zygomatic arches and widely flared nasal bones. It is known from southern, eastern and south-western areas of Australia where it inhabited dry sclerophyll forests, near waterways, and fed on browse including roots and tubers. In these forests also lived *Phascolarctos yorkensis*, a Plio-Pleistocene koala that was about a third larger than the modern Koala *P. cinereus*, making it the largest arboreal marsupial of its time. During the Pleistocene, these two koala species appear to have co-existed in some areas and it remains unclear why the larger one ultimately became extinct. In the distance, short-faced kangaroos *Procoptodon rapha* browse on leaves from trees and shrubs, while *Macropus giganteus titan*, a megafaunal relative of the modern Eastern Grey Kangaroo *M. giganteus* grazes on grasses nearby. The Eastern Bearded Dragon *Pogona barbata* is an agamid lizard that can still be seen today in wooded parts of Australia. ●

AGE

Pleistocene, 2.6 million to about 11 000 years ago.

LOCALITY

Wellington Caves Reserve, Wellington, New South Wales.

ENVIRONMENT

Open sclerophyll forests and woodlands with a grassy/shrub understorey, similar to the vegetation that currently covers the hills in which the Wellington Caves occur. However, it is possible that *Zygomaturus trilobus* needed wetter terrains or even swamps which may have occurred in lower areas of the same region associated with the ancestral Bell River.

REFERENCES

28, 55, 92, 93, 223, 324

Wellington Caves, Scene 2, New South Wales

WHEN MOST OLDER AUSTRALIANS THINK OF
kangaroos, they tend to think of 'Skippy', the name
of a putatively intelligent, grass-munching Eastern
Grey Kangaroo *Macropus giganteus* that featured
in an Australian television show between 1968 and
1970. Skippy's hands were often shown 'untying'
knots to save someone being abused by human
villains. There was one group of extinct kangaroos
that would save you from harm only if they could
then eat you: the carnivorous to omnivorous
propleopine hypsiprymnodontid kangaroos. Earlier
forms like *Ekaltadeta ima* from the Miocene of
Riversleigh were almost certainly dominantly
carnivorous. Their younger larger descendants, the
Pleistocene species of *Propleopus* and Pliocene
species of *Jackmahoneya*, may have been more like
omnivorous bears, capable and keen to eat animals
but perhaps more commonly dining on fruits,
flowers or soft leaves. The most common of these
was *P. oscillans*, which is known from several fossil
deposits in the eastern states of Australia. This
distant relative of the living omnivorous but much
smaller Musky Rat-kangaroo *Hypsiprymnodon
moschatus* which inhabits the rainforests of north-
east Queensland may have been about 70 kg in
weight, about the size of an Eastern Grey Kangaroo.
But as well as having a far more catholic diet, like
other propleopines it probably galloped rather
than hopped – altogether a very un-Skippy-like
kangaroo. The fowl dinner it is shown dining on is
the Australian Brush-turkey *Alectura lathami*, which
would have occupied the same habitat as this
kangaroo and often graced its table. ●

AGE
Pleistocene, 2.6 million to perhaps 35 000
years ago.

LOCALITY
Although this species was never common,
perhaps reflecting its role as an opportunistic
carnivore in Australia's Pleistocene
ecosystems, *P. oscillans* was reasonably
widespread in eastern Queensland (mainly
the Darling Downs), New South Wales
(e.g. Wellington Caves as well as Lake
Menindee), Victoria and South Australia.

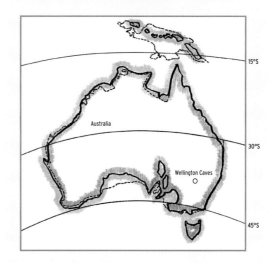

ENVIRONMENT
Most of the deposits in which this species is
known to occur would have accumulated in
open forest or woodlands rather than closed
forests.

REFERENCES
51, 92, 93, 296, 332

Wellington Caves, Scene 3, New South Wales

SINCE 1830, THE PLEISTOCENE MAMMALS OF Wellington Caves have had an extremely important role in developing early understanding about the globally unique nature of Australia's ancient mammals as well as in the initial development of Darwin's realisation about the reality of evolution. After visiting South America and seeing that the ancient mammals of that land most closely resembled that continent's living types, the biblical claim about a global flood began to seem improbable. Subsequent recognition that the fossil mammals from Wellington Caves included wombats, kangaroos and possums rather than rhinoceroses, elephants and lions finally convinced Darwin that animals had evolved on each continent from pre-existing kinds unique to those lands. That said, when Sir Richard Owen first saw specimens of the animal he named *Thylacoleo carnifex*, which came to be called the Marsupial Lion, he realised it was in many ways 'lion-like', albeit a decidedly unique marsupial version that had nothing to do with placental lions other than the fact that they both were ferocious carnivores. Weighing up to 160 kg and equipped with a massive thumb claw, powerful dagger-like lower incisors, lethally sharp and enormous carnassial third premolars, and powerful arms, there were no megafaunal mammals in Australia immune to becoming this globally most specialised carnivorous mammal's dinner.

Among the millions of fossil bones and teeth jumbled together like a Christmas pudding in the reddish cave deposits exposed in the Phosphate Mine at Wellington are those of some of the largest kangaroos that have ever terrorised a leaf on this continent. Among them was the gigantic 'wallaby' *Protemnodon brehus*, which occurred in all states of Australia except Tasmania. While there were many other species of this genus spread between Tasmania and New Guinea, at perhaps 110 kg this was one of the largest. Like other species of the same genus, it had very long shearing premolars for cutting twigs and tough leaves. As a browser it would have spurned the rapidly expanding grasslands, preferring the leaves of shrubs and the low branches of trees, from where it would have kept a wary eye out for hungry Marsupial Lions. ●

AGE

Mid to late Pleistocene, about 1 million–500 000 years ago.

LOCALITY

Phosphate Mine, Wellington Caves, New South Wales.

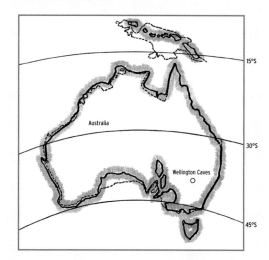

ENVIRONMENT

The vegetation surrounding the Wellington Caves would have included sclerophyll forests, shrublands and grasslands perhaps similar to the vegetation that covers and surrounds these hills today. Depending on the particular part of the Pleistocene that hosted these animals, the overall climate of the area would have been semi-arid to Mediterranean.

REFERENCES

28, 92, 93, 170, 223

The giant lizard Megalania, eastern Australia

PLEISTOCENE

AN EASTERN GREY KANGAROO LEAPS TO ESCAPE ambush by the world's largest known terrestrial lizard, Megalania *Varanus priscus*, Australia's giant extinct monitor lizard. Growing to 5.5 m long or more, and up to 500 kg in weight, this gigantic lizard was at least twice the size of its living relative, the Komodo Dragon *Varanus komodoensis* of Flores and nearby islands. During the Pleistocene, it was Australia's largest terrestrial carnivore. It could have taken down the largest kangaroos and probably even the rhino-sized Diprotodon *Diprotodon optatum*, which it would have torn to pieces using its very large claws and recurved serrated teeth. It was probably also venomous, like closely related living varanids which

use a potent hemotoxin released by glands in the mouth that acts as an anticoagulant and greatly increases bleeding in wounded prey. Speculation about how living Komodo Dragons kill large prey led to the initial suggestion that bacteria in their mouths caused lethal infections after they bit the legs of the intended prey. When it was discovered that varanids have venom, that was suspected of being the way they subdued prey. But more recent studies suggest that the Komodo Dragons' primary method of killing large ungulates such as cattle is the action of their jagged teeth, which sever blood vessels in the legs and lead to death from blood loss. The Komodo Dragons only have to follow the weakening prey until it drops, unable to defend itself. This may well be the same strategy used by Megalania to safely bring down potential gigantic meals like Diprotodon. In the Pleistocene, Megalania lived alongside smaller varanids, such as the living Perentie and Lace Monitor. But it was not the only giant lizard roaming Australia at that time – fossils from Mount Etna caves in Rockhampton, Queensland indicate that the extant Komodo Dragon also once lived and probably first evolved in Australia. Another very large monitor species is known from fossil deposits in the Lake Eyre Basin of central Australia. ●

AGE
Pleistocene until at least 50 000 years ago.

LOCALITY
Megalania was widely distributed but seemingly very rare across much of eastern Australia. Although skeletons are unknown, teeth, vertebrae and other bones have been found in New South Wales, South Australia, Victoria and Queensland, particularly in the Darling Downs region. It is not yet known from Tasmania, Western Australia or New Guinea.

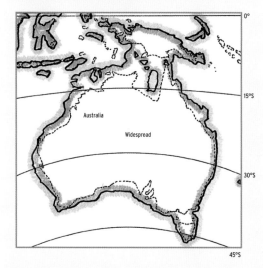

ENVIRONMENT
During the Pleistocene, Megalania probably lived in open woodlands and grasslands across at least the eastern half of mainland Australia.

REFERENCES
101, 175, 258, 314

Thylacoleo Caves, Western Australia

ONE OF THE MORE REMARKABLE MODERN discoveries of Pleistocene vertebrates was reported by Gavin Prideaux and colleagues in 2007, from three closely associated cave deposits called Thylacoleo Caves, on the Western Australian Nullarbor Plain. At last count, 68 mammal species have been found including many new species; for example, eight new kangaroo species among the total of 23 recovered from the deposits. Among the new species was a giant tree kangaroo *Bohra illuminata*. As the first known skull and skeleton of an extinct member of this fascinating group of

kangaroos, they exhibit features that demonstrate these were indeed arboreal kangaroos. But at around 40 kg, these tree-climbing roos were twice the size of any of the modern tree kangaroos (*Dendrolagus* sp.) living in the rainforests of New Guinea and Australia. There are also features that demonstrate older evolutionary relationships to rock wallabies (*Petrogale* sp.) some of which have been known to climb trees as well as rocky hillsides. Other kangaroos in these deposits include the very strange *Congruus kitcheneri* which had first been found in a fossil deposit in Western

Australia. Other new species of *Congruus* in the Thylacoleo Caves deposits include one that may have had horn-like growths projecting from the skull over the orbits. Among the long list of extinct megafaunal species in these deposits is the Giant Wombat *Phascolonus gigas* which, if it burrowed like its modern relatives, could have been excavating tunnel systems large enough to harbour small cows. Skeletons of the extraordinary carnivorous Marsupial Lion *Thylacoleo carnifex* in these deposits are the best preserved of any so far found. On balance, they confirm that this was the most specialised mammalian carnivore that has evolved anywhere in the world. But they also reveal a uniquely upwardly bent tail that supported this predator as it leaned back on its haunches, thereby freeing its powerful arms to kill or dismember its prey. ●

AGE
Middle Pleistocene, with many of the fossils apparently accumulated 400 000-230 000 years ago.

LOCALITY
Thylacoleo Caves (Leaena's Breath Cave, Flight Star Cave, Last Tree Cave), Nullarbor Plain, Western Australia.

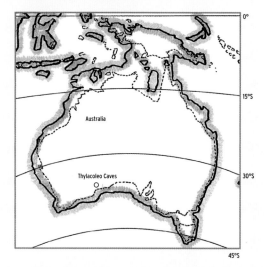

ENVIRONMENT
Dry, relatively open sclerophyll/woodland and shrubland environment, similar to many areas of central Australia today. Although they are not in the area now, some relatively large trees must have been present at this time to accommodate hollow-nesting parrots and two different species of giant tree kangaroos.

REFERENCES
109, 243, 318, 319, 320, 413

Naracoorte Caves, Scene 1, South Australia

PLEISTOCENE

BECAUSE OF THEIR EXTRAORDINARY RECORD OF Pleistocene vertebrates, Naracoorte Caves were listed with Riversleigh in 1994 as a serial World Heritage property called Australian Fossil Mammal Sites (Riversleigh/Naracoorte). While Pleistocene megafaunal fossils are also known from Riversleigh, those from Naracoorte are far more diverse and better preserved. The deposits, which have been found in 26 different caves including the fossil-rich Blanche and Victoria Caves, contain many of the extinct Pleistocene vertebrate species that were once widespread throughout southern and eastern Australia. For example, the browsing Short-faced Kangaroo *Simosthenurus occidentalis* (middle right), first discovered in Mammoth Cave in Western

Australia, the perhaps hippo-like diprotodontid *Zygomaturus trilobus* (lower right), first discovered in Tasmania, and the hypercarnivorous *Sarcophilus laniarius* (lower left among logs), first discovered in the Wellington Caves in New South Wales, have been found together in the Naracoorte deposits. Among the many other extinct species that have been found here are the giant long-beaked and probably worm-eating echidna *Megalibgwilia ramsayi* (lower middle) and the archaic wombat *Warendja wakefieldi* (lower left). This wombat is a late Pleistocene survivor of the earliest lineage to develop rootless teeth, an adaptation which Riversleigh's fossil taxa (e.g. *W. encorensis*) have demonstrated evolved over the last 15 million years in response to the increasingly abrasive vegetation of a drying Australia. Among the many other fossilised carnivores in Naracoorte's limestone treasure houses are the Thylacine *Thylacinus cynocephalus* (middle left), which survived on the mainland until about 4000 years ago, possibly going extinct because of competition with introduced Dingoes, and the at least partly arboreal Marsupial Lion *Thylacoleo carnifex*, shown here surveying from above the possible components of the day's menu, above a Diprotodon *Diprotodon optatum*. Although a hypercarnivore, which are normally uncommon in ecosystems, there are surprisingly high numbers of Marsupial Lion specimens in the Naracoorte Caves deposits. This may be because the scent of dead or dying animals that had fallen into the caves attracted a disproportionate number of carnivores. ●

AGE
Late Pleistocene, most fossils accumulated about 400 000–200 000 years ago.

LOCALITY
Naracoorte Caves, south-eastern South Australia.

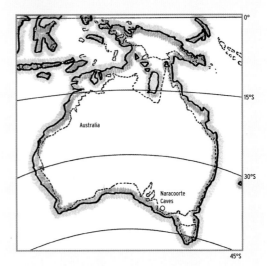

ENVIRONMENT
Tall eucalypt forests with a thick understorey of hard-leaved shrubs alternating over time and space with woodlands, grasslands and swampy sedge lands. However, during most of the late Pleistocene when the Naracoorte fossils were accumulating, most of the area would have been open eucalypt woodlands.

REFERENCES
128, 232, 233, 270, 413

Naracoorte Caves, Scene 2, South Australia

PLEISTOCENE

OF ALL THE KINDS OF SNAKES KNOWN FROM Australia, the most fascinating and mysterious are the extinct madtsoiids. Madtsoiids once thrived in South America, Africa, Madagascar, India, Australia and even some areas of southern Europe, from the late Cretaceous to late Pleistocene. However, in most areas of the world except Australia they were gone by the end of the Eocene. In Australia, they survived until the late Pleistocene. Decline of the group within Australia may have been the result of steady drying out of the continent. Madtsoiids are among the most 'primitive' snakes known, with the family falling outside all other still-surviving snake families. Because they retain many archaic skull features, they are a primary focus of research into the origin of snakes as a whole. While uncertain, it would appear that some, such as the madtsoiids from Riversleigh, were at least semi-aquatic, possibly having lifestyles similar to modern anacondas. Others such as *Wonambi naracoortensis*, which occupied drier environments in southern Australia, are more likely to have been semi-fossorial than semi-aquatic. Probably they were non-venomous constrictors, although there is no certainty about these probabilities either. *Wonambi naracoortensis* may have survived long enough to overlap in time with the first humans in Australia and it has been suggested that early sightings of this large 6 m snake, which probably had iridescent scales, might even have been the origin of the legend of the Rainbow Serpent. *Wonambi* is an Aboriginal word for a Dreamtime legend about a Rainbow Serpent. ●

AGE

Late Pleistocene.

LOCALITY

Naracoorte Caves, south-eastern South Australia.

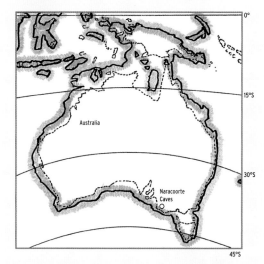

ENVIRONMENT

The habitat around the Naracoorte Caves in the late Pleistocene would have been dry sclerophyll forest and woodlands.

REFERENCES

33, 263, 285, 333, 348

Texas Caves, Scene 1, Queensland

AS CURRENTLY UNDERSTOOD, THERE ARE 12 species of extinct short-faced browsing kangaroos placed in the genus *Procoptodon*. They exhibited many of the morphological extremes of kangaroos, including gigantism with some species up to 2 m tall, weights of almost 250 kg and loss of all toes except the fourth toe of each hind foot. Even their hands were specialised, such that two fingers of each hand were extremely elongated, presumably adapted for grasping branches high above the reach of all other roos. They are known only from Pleistocene deposits. A series of caves near Texas, south-eastern Queensland, contain Pleistocene deposits that were excavated in a rush by the Queensland Museum in 1975 just before they went under water when the Glenlyon Dam was built on Pike Creek in 1976. A fossil-rich limestone found in one cave known as The Joint produced, besides the first described dagger-toothed mekosuchine crocodile from Australia, a single tooth (P[3]) of a giant macropodid, *Procoptodon texasensis* (which may or may not be the same species as *P. rapha*). This find was extremely fortunate considering that at the time of the excavation the cave was filled with CO_2 which nearly overcame the palaeontologist (Mike Archer), who struggled to remain awake while using a hammer and chisel to gouge out blocks of the breccia. Major Mitchell's Cockatoo *Lophochroa leadbeateri* would have been present in the dry sclerophyll forest that surrounded The Joint as it accumulated the teeth and bones of animals that stumbled into the crevasse above. ●

AGE

Pleistocene, more than 292 000 years ago (precise age unknown).

LOCALITY

Texas Caves, south-east Queensland.

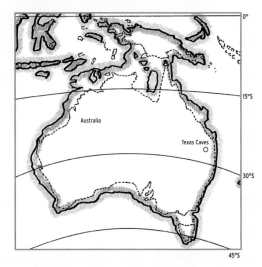

ENVIRONMENT

The fauna from The Joint deposit suggests that the palaeoenvironment in this region of south-eastern Queensland in the Pleistocene was dry to wet sclerophyll forest. Two species, a pademelon and bandicoot (*Thylogale* sp. and *Perameles nasuta*), however, suggest the possibility that there were patches of rainforest nearby. The presence of a mekosuchine crocodile does not necessarily indicate water, given that some of these crocodiles are regarded to have been at least in part terrestrial.

REFERENCES

11, 168, 170, 318

Texas Caves, Scene 2, Queensland

PLEISTOCENE

WHEN NEWS CAME THAT THE GLENLYON DAM was being built and that it would back up Pike Creek in south-eastern Queensland, the Texas Caves were suddenly in imminent danger of being flooded forever. Mike Archer mounted a Queensland Museum expedition to excavate as many of the fossil deposits as possible, and to survey the modern biota of the area. The fossil focus was on Russenden Cave and a fissure deposit known as The Joint. Fossils recovered from Russenden Cave included thousands of bones representing a wide range of animals, some of which had gone extinct in the region and some of which still survive. Among the extinct forms was a fascinating new species of *Antechinus* later named *A. puteus* by Steve Van Dyck; the fossil deposit in The Joint provided even more unexpected creatures. Although the precise age of these Texas Cave deposits is uncertain, some species found in The Joint deposit, such as the giant short-faced kangaroo *Procoptodon texasensis*, indicate a Pleistocene age that has been determined to be greater than 292 000 years old. Other species found included a wombat that is probably conspecific with the Common Wombat *Vombatus ursinus*, a large marsupial still living in this region of Queensland. But not all the Texas Cave discoveries represented animals similar to others already known. A maxilla of a reptile that emerged from The Joint deposit was like nothing ever seen before. While it was clear that it did not represent a lizard – because the teeth were in sockets in the bone rather than attached to the sides of the bone – it nevertheless had blade-like recurved serrated teeth like those of a goanna. This was the first discovery of what eventually proved to be the highly distinctive subfamily of deep-headed semi-terrestrial crocodiles, the Mekosuchinae. This one was eventually named *Quinkana fortirostrum*. ●

AGE

Pleistocene; The Joint, more than 292 000 years ago (precise age unknown); Russenden Cave, 55 000 years ago.

LOCALITY

Unnamed fossil deposits, Russenden Cave and The Joint, Texas Caves, Pike Creek area, south-east Queensland.

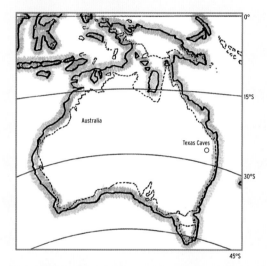

ENVIRONMENT

The faunas suggest the area in the region of the caves supported relatively dry sclerophyll forest with grasslands, possibly with pockets of wet sclerophyll forest and/or rainforest in adjacent areas.

REFERENCES

11, 223, 317, 401

Lord Howe Island meiolaniid turtle

AT GROUND LEVEL, GIANT LUMBERING TURTLES own this place. The Lord Howe Island meiolaniid *Meiolania platyceps*, with shells exceeding 1 m in length, had sheep-like horns protruding from their heads and long spiked tails ending in a large bony club. But despite their 'battle armour', these animals were no threat to other animals – they were plant-eaters. Their armour was presumably an evolutionary holdover from years and places past when their ancestors in Australia and other areas of Gondwana needed to protect themselves from a range of fearsome predators. At some

point, a million or more years ago, adventurous or incautious ancestors of these Lord Howe Island turtles floated across the Tasman to land on Lord Howe Island. Presumably this proclivity for frolicking around the shores of Australia meant that many other less fortunate turtle travellers also drifted away, only to become a shark's dinner or a lump on the bottom of the ocean. The first successful immigrants to Lord Howe Island would have encountered avian denizens (terrestrial mammals do not appear to have reached Lord Howe Island) including a couple of rails, the most striking of which was the White Swamphen *Porphyrio alba* (left), a close relative of the Purple Swamphen *P. melanotus* of Australia and New Zealand. Like its relatives, these Lord Howe species would have fed on a range of plants as well as any tasty invertebrates. It is entirely possible that these omnivorous groundbirds also indulged, whenever the opportunity arose, on baby *Meiolania platyceps* turtles when they hatched, in the same way other rails and herons do today on islands where sea turtles breed. Abundant pigeons *Columba vitiensis godmanae* (top left) and parakeets *Cyanoramphus novaezelandiae subflavescens* (bottom right) also lived in these forests. The pigeons probably spent most of their time feeding in the forest canopy, but the parakeets would have made regular forays to the forest floor to feed on ground plants and seeds. Exactly when this turtle-dominated ecosystem collapsed on Lord Howe we do not know, but in Vanuatu some of the last meiolaniids appear to have met their end in the ovens of the first Vanuatuans about 2500 years ago. ●

AGE

Late Pleistocene to Holocene, about 100 000–1000 years ago.

LOCALITY

Lord Howe Island, Tasman Sea.

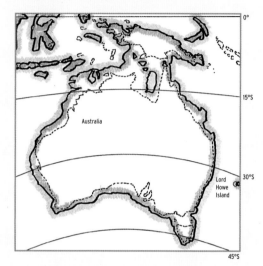

ENVIRONMENT

In the lowland subtropical forest that clothes the dunes at the back of what is now Ned's Beach, Kentia palms (*Howea* sp.) form thick stands and pandanus groves fill the spaces between these and the sea The calcareous sands that form the dunes near the beach appear to have been able to entomb any horned turtles that were unlucky enough to roll over or be trapped in depressions in the ground.

REFERENCES

130, 131, 133, 284, 419

Mammoth Cave, Western Australia

MAMMOTH CAVE, IN THE DENSE TALL EUCALYPT forests south-west of Margaret River, is a most curious place. Fossil bone deposits containing at least 34 different kinds of vertebrates including the now extinct Thylacine *Thylacinus cynocephalus* were discovered there as early as 1904. Most of the animals in the deposit appear to have fallen into the cave through a now-blocked solution pipe in the roof. Possible pre-European fossil collecting in Mammoth Cave has been suggested as an explanation of the 'charms' valued by Indigenous Australians in the Kimberley area of north-western

Australia. These include a third upper premolar of the hippopotamus-like diprotodontid *Zygomaturus trilobus*, set in spinifex gum and attached to a string made of hair. Other Kimberley charms included isolated teeth of the medium-sized sthenurine kangaroo *Simosthenurus browneorum*. Both species are common in the fossil deposits of Mammoth Cave but not known from northern Australia. Although trade in fossils up and down the Western Australia coast is possible, it is also possible – although perhaps less plausible – that there is an as yet undiscovered fossil deposit with these species somewhere in the Kimberley region.

One of the most intriguing fossils from Mammoth Cave, collected sometime between 1909 and 1915 by Ludwig Glauert, who was excavating by candlelight on behalf of the Western Australian Museum, was an extinct giant long-beaked echidna *Murrayglossus hacketti*, the largest monotreme known. Several kinds of long-beaked echidnas are known from Pleistocene fossil deposits in Australia, but today species of this genus are confined to New Guinea and adjacent islands. A recent claim that the Western Long-beaked Echidna *Zaglossus bruijnii*, known today only from New Guinea, survived into modern times in north-western Australia may be based on a label mix-up in the Natural History Museum in London. However, why the long-beaked echidnas like *M. hacketti* died out in the late Pleistocene, leaving in Australia only the common Short-beaked Echidna *Tachyglossus aculeatus*, is a mystery. ●

AGE

Late Pleistocene (precise age uncertain). Radiometric dates of items recovered from remnants of the Glauert Deposit still adhering to the cave wall have provided dates from about 55 000–44 000 years ago.

LOCALITY

Mammoth Cave, Leeuwin-Naturaliste National Park, south-western Western Australia.

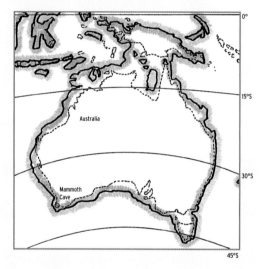

ENVIRONMENT

The list of species known from the Mammoth Cave deposit, in addition to the extinct forms, includes Koalas, rock-wallabies and other modern types that still abound in forested areas of the structural kind that surrounds Mammoth Cave today. Hence, it is possible that the environment at the time this deposit accumulated may have been relatively similar to that which still occurs today.

REFERENCES

5, 17, 147, 167, 247, 248

Mowbray, Victoria
PLEISTOCENE

MOWBRAY SWAMP IN NORTH-WEST TASMANIA has produced fossils of two of the most common Pleistocene marsupials that also commonly occur in continentally marginal sites around Australia: the zygomaturine diprotodontid *Zygomaturus trilobus* and the macropodine kangaroo *Protemnodon anak*. One of the two skeletons of *Z. trilobus* recovered from this deposit was first described by colonial palaeontologists, H.H. Scott of the Queen Victoria Museum in Launceston and C. Lord of the Tasmanian Museum in Hobart. It was a strange animal, possibly 500 kg in weight, with a wide foreshortened turned-up snout that may have supported some kind of a fleshy mobile trunk. It may have been a semi-aquatic herbivore like the Hippopotamus. It had hollow sinuses in the bones of the head that enabled the skull's outer surface to be sufficiently large for attachment of its massive chewing muscles. These 'air pockets' took up almost 25% of the total volume of the head, which is why diprotodontids exhibiting this sort of pneumatisation of their cranial bones have been called 'air heads'. *Protemnodon anak* has been described as a huge 'wallaby'. Its relationships to other kangaroos have been controversial, with recent analysis of the skulls suggesting it falls outside of all the macropodine kangaroos including the wallabies and larger species. However, recent remarkable analysis of DNA recovered from a Tasmanian specimen that was about 45 000 years old suggests that the species of *Protemnodon* were most closely related just to kangaroos of the genus *Macropus*, hence tucking in much further up the family tree. They had very elongate skulls. Their long narrow premolars and low-crowned molars suggest that they were, like the species of *Zygomaturus*, browsers rather than grazers. Although most of the extinct megafaunal species were gone from Tasmania before humans arrived, perhaps 45 000 years ago, it is possible that these two species were still present. Nevertheless, analysis of the bones of all extinct Tasmanian megafaunal species has revealed no evidence that humans had a role in their extinction. ●

AGE

Late Pleistocene, probably greater than 52 000 years ago. Efforts to date the deposit have led to the conclusion that the dates obtained are unreliable because of contamination by modern carbon sources.

LOCALITY

Mowbray Swamp, a 10 km wide deposit in the north-western corner of Tasmania, south-west of Smithton.

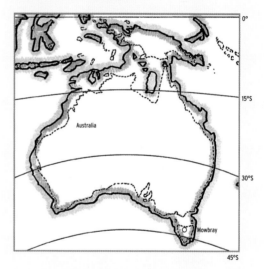

ENVIRONMENT

Pollen diagrams suggest the presence of forest trees, tree ferns and abundant shrubs during the approximate time when these species lived in the Mowbray Swamp area.

REFERENCES

94, 145, 146, 214, 353

Callabonna, South Australia
PLEISTOCENE

ONE OF THE MOST AMAZING THINGS ABOUT LAKE Callabonna is a picture taken years ago of palaeontologist Richard Tedford squatting next to a ghostly series of enormous footprints that disappear off into the distance. They are the footprints of *Diprotodon optatum*, the largest marsupial that ever lived, and perhaps record the journey of a doomed giant searching for disappearing water during an ancient drought. If that is what happened, this individual was not struggling alone. Entombed in Callabonna's late Pleistocene clays are the skeletons of many

different kinds of extinct as well as still-living animals that became mired in the treacherous muds of the drying lake. Extinct species include short-faced kangaroos (the giant *Sthenurus stirlingi* [upper left] and smaller *S. andersoni* [middle left]), cow-sized wombats *Phascolonus gigas* (upper right), the last survivor of the once diverse mirihungs or thunderbirds *Genyornis newtoni* and many other fascinating creatures. The still-surviving fish-eating Shag *Phalacrocorax melanoleucus* shown in the diorama, along with many other species that survived the sequence of late Pleistocene extinctions, would have been part of a vast complex ecosystem when Lake Callabonna was full of water. Even today, on the rare occasions when the lake does fill with water, vast numbers of waterbirds appear within days and repopulate the arid area until the waters gradually disappear once again. Fossils in the lake's deposits were first reported in 1882. Although there have been many expeditions to collect the fossils, the most famous were those in 1893 carried out by Sir Edward C. Stirling and Amandus H.C. Zeitz of the South Australian Museum who collected some of the best skeletons known of *D. optatum* and *G. newtoni*. This led to establishment of the lake as a Fossil Reserve in 1901, one of the first in Australia. More recent expeditions, including those led by Trevor Worthy of Flinders University, have made even more extraordinary discoveries, confirming the wisdom of the 1997 decision to place Lake Callabonna on the South Australian Heritage Register. ●

AGE

Late Pleistocene.

LOCALITY

Although the modern dry salty Lake Callabonna in central northern South Australia is 160 km², it was probably much larger when the fossil deposits were accumulating. Most of the fossil deposits have been found near the north-eastern edge of the lake in an area that has become the Lake Callabonna Fossil Reserve. The lake surface where the fossil beds are exposed range -2 to -4 m below sea level.

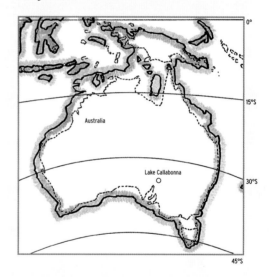

ENVIRONMENT

Callitris (conifer) cones and other plant materials found in the deposit alongside the fossil bones indicate that the area evidently supported an open savannah woodland with at least some trees as well as shrubs.

REFERENCES

299, 372, 381, 415

Wyandotte Station, Queensland

PLEISTOCENE

AUSTRALIA WAS ONCE HOME TO A FAMILY (Meiolaniidae) of large herbivorous terrestrial turtles characterised by having highly distinctive armoured heads and tails, adorned with horns and tail clubs. The largest of these meiolaniid turtles was the Wyandotte Giant Horned Turtle *Meiolania* sp. cf. *M. platyceps* of northern Queensland, which is estimated to have a carapace (upper shell) some 2 m long. Fossils of this giant turtle include three cow-like horn cores and a tail vertebra recovered from Pleistocene sediments along Wyandotte Creek near Greenvale. This species is distinct from the giant *Ninjemys oweni*, indicating that there were two species of giant armoured turtles living

in Queensland during the late Pleistocene. The Wyandotte horned turtle was more similar to the much smaller Lord Howe Island's *Meiolania platyceps*. Apart from mainland Australia and Lord Howe Island, meiolaniid turtles once also lived in South America, New Caledonia, Vanuatu, Fiji and possibly New Zealand, suggesting that they are in fact a Gondwanan group. Their occurrence in places like Lord Howe and Vanuatu indicates that, although essentially terrestrial, they must have been able to float long distances in ocean waters, enabling them to colonise remote islands. The family is very old, possibly branching off the rest of the turtle family tree before the split between the two modern turtle lineages, the cryptodires (which pull their head straight back into the shell) and pleurodires (which fold their neck sideways under the shell). The oldest known undoubted meiolaniids in Australia come from the Eocene Rundle Shale in Queensland. The Wyandotte Giant Horned Turtle is shown here with a Bridled Nail-tail Wallaby *Onychogalea fraenata*, an endangered small grazing kangaroo that survives today in Queensland in small, isolated populations around Emerald. These meiolaniid bones, along with those of other living and extinct species, have been found eroding out of Pleistocene sediments over a 10 km stretch along Wyandotte Creek. Other species found here include the giant monitor lizard Megalania *Varanus priscus*, a madtsoiid snake (*Wonambi* sp.), the bizarre Marsupial Tapir *Palorchestes azael*, Diprotodon *D. optatum* and the Giant Wombat *Phascolonus gigas*, as well as extant species of ducks, geese, bandicoots, antechinuses, rodents and elapid snakes. ●

AGE

Late Pleistocene, about 43 000 years ago.

LOCALITY

Clays, sands and gravels exposed along Wyandotte Creek, northern Queensland.

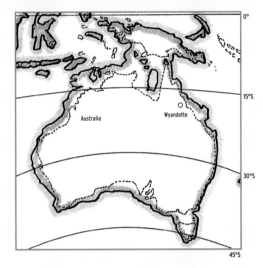

ENVIRONMENT

During the late Pleistocene, this area probably would have been incised by braided streams surrounded by woodlands and grasslands.

REFERENCES

133, 136, 242, 305, 312, 371

Pureni, Papua New Guinea
PLEISTOCENE

VERY FEW OF AUSTRALIA'S DIPROTODONTID
marsupials were hippo- or rhino-like other
than some of the largest kinds, like species of
Zygomaturus and *Diprotodon*. As more skeletal
remains of the smaller kinds turn up, there
has been growing awareness that some had
significantly different lifestyles. This may have been
the case, for example, for *Hulitherium tomasettii*,
a browsing zygomaturine diprotodontid from the
late Pleistocene of New Guinea. Although what
is known of its skull suggests some odd features
such as the arched roof of the mouth, its femur is
a stand-out oddity. The ball that articulates with
the pelvis projects up above the shaft rather than
sticking out to the side, suggesting that it had
highly unusual, very extensive mobility. Its humerus
similarly suggests unusually high mobility for the
arm. It is possible that this rainforest browser may
have been partially arboreal, like Himalayan Sun
Bears or even Pandas. Despite being much larger,
with an estimated weight of 75–200 kg and a
body length of perhaps 2 m, in terms of lifestyle
H. tomasetti may have been similar to the arboreal
Miocene species of *Nimbadon* and *Silvabestius*
from Riversleigh. Because its closest relationships
appear to be with the equally distinctive
Pleistocene New Guinean zygomaturine *Maokopia
ronaldi*, it has been suggested that both are the
products of a distinctive regional radiation that
followed the arrival of the first diprotodontids into
New Guinea. These first, perhaps late Miocene or
early Pliocene immigrants were perhaps similar to
species of *Plaisiodon* or *Zygomaturus*. The Ribbon-
tailed Astrapia *Astrapia mayeri*, whose tail feathers
are relatively longer than those of any other bird in
the world, is a modern bird of paradise restricted to
subalpine forests in the central highlands. Although
not known from the Pureni fossil deposit that
contains *H. tomasetti*, it may well have danced in
the same tree branches that strained under the
weight of this now extinct, possibly semi-arboreal
marsupial. ●

AGE

Pleistocene, about 38 600 years old.

LOCALITY

Pureni, Southern Highlands, Papua
New Guinea.

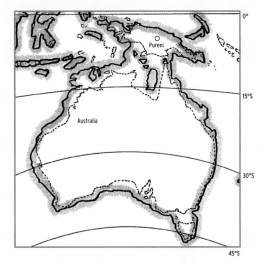

ENVIRONMENT

The Pureni deposit spans a series of habitat
changes in this area from open water
surrounded by forest to a grassy swamp and
ultimately a cool mossy upland bog forest.
Throughout the interval of deposition, the
temperature was colder than the area is at
present.

REFERENCES

115, 122, 223

Riversleigh, Scene 12, Queensland

PETER SCHOUTEN 2013

TERRACE SITE IS A PLEISTOCENE RIVER DEPOSIT that was found exposed in the banks adjacent to the modern Gregory River that runs through Riversleigh Station in north-western Queensland. The fossil deposit itself is an unconsolidated bed of gravel that was being transported by an ancestral anabranch of the Gregory. A wide range of animals that lived in the river as well as those in the woodlands along its banks ended up contributing bones to this deposit. Lumps of charcoal near the base of this deposit were used to obtain a radiocarbon date of 23 900 years

ago, with a 'standard error' that meant there was a 95% probability that the actual date was between 21 200 and 28 000 years before the present. This is important, because this would be one of the youngest dates in Australia for *Diprotodon optatum* and *Palorchestes azael*. In this reconstruction, *P. azael* is shown in the water, possibly eating nutritious waterlily tubers that it has clawed up out of the river bed. But what it actually ate, and how it used its enormous, clawed forelimbs, remain controversial. The snout of a powerful 5 m crocodile *Paludirex gracilis* was also recovered from this same deposit. Agile Wallabies *Macropus agilis* were then and remain today one of the most common mammals found along the banks of the Gregory River. *Elseya lavarackorum* is a most curious member of this community. When first found, the large shells were clearly from a turtle that had not previously been scientifically described. It was named by Arthur White and Mike Archer after Jim and Sue Lavarack, the Riversleigh volunteers who found and spent many long days in the quarry carefully excavating the large carapace of this turtle. A short while later, Arthur was swimming in the nearby Lawn Hill Creek when he bumped into a strange turtle. When he caught it, to his amazement it appeared to be similar to the one they had just named on the basis of the Terrace Site specimen. Once presumed extinct, this 'Lazarus turtle' appeared to have miraculously come back to life! However, more recent research has demonstrated that the living form, which is now known as the Gulf Snapping Turtle, is in fact another new species, *Elseya oneiros* – making the point that we still do not know enough about Australia's living animals! Ironically, a second less-preserved fossil turtle found in the Terrace Site deposit turns out to be the same or a very close relative of *E. oneiros*, but it wasn't given a formal name when discovered. In the background of the painting, people are shown foraging along the bank near the largest marsupial that ever lived, Diprotodon *D. optatum*. There is no hard evidence that these marsupials were hunted, let alone driven to extinction, by humans. What evidence there is suggests that humans and these giant marsupials overlapped for a long time before climate change triggered the extinction of these huge herbivores. Also possibly present at this time was another extinct giant of the Pleistocene, the flightless dromornithid bird *Genyornis newtoni*. While it may have been here, specimens were said to have been seen but not found in immediately adjacent similar riverine deposits – a challenge for future work at this important site. ●

AGE

Late Pleistocene, between about 28 000 and 21 000 years ago.

LOCALITY

Terrace Site, Gregory River deposits, Riversleigh World Heritage Area, north-west Queensland.

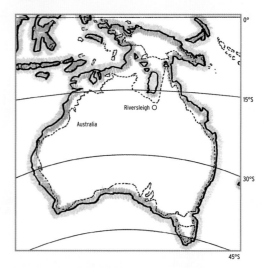

ENVIRONMENT

The Pleistocene was marked by rapid and severe changes in climate and environmental conditions. Like the rest of the world, Australia experienced extreme climatic conditions as the Arctic and Antarctic polar ice caps repeatedly grew and retreated, resulting in a constantly changing pattern of forests, grasslands and deserts because of consequent fluctuations in temperature and rainfall. Despite cyclical fluctuations, the overall trend throughout the Pleistocene was towards increased dryness.

REFERENCES

51, 91, 311, 418

WHILE THE LAST OF THE LATE CENOZOIC
diprotodontids in Australia such as Diprotodon *Diprotodon optatum* were 2–3 t in weight, others in New Guinea were very different. At about 100 kg, *Maokopia ronaldi* was a relatively small zygomaturine diprotodontid. Although it is known only from cranial remains, these suggest it was most closely related to the slightly larger *Hulitherium tomasettii*. Both of these late surviving New Guinean members of this family may have been semi-arboreal, like Asian Sun Bears. Alternatively, it has been suggested that they may have been more Panda-like in their behaviour as well as appearance. Until more of the skeletons of these marsupials are known, we can only begin to guess at their lifestyle, although in cranial size and shape they were clearly unlike other late Cenozoic diprotodontids. Their late survival, in spite of probable overlap with humans for at least 40 000 years, suggests their ultimate extinction may have had complex causes. Other animals recovered from this deposit include small birds, marsupials, rodents and bats, some possibly brought into the cave as prey by carnivorous owls. The only other large species found was the extinct wallaby *Protemnodon hopei*. This browsing kangaroo may have been the last surviving member of the genus *Protemnodon*, given that all others were probably extinct on the Australian mainland by 35 000 years ago. The Snow Mountains Robin *Petroica archboldi* shown here, which is endemic to mountains in Papua and Indonesia, occurs from about 3850–4150 m above sea level, well above the current snowline. Although not yet known to be present in the Kelangurr Cave deposit, it is possible that it co-occurred with *M. ronaldi* in the same type of high-altitude grasslands and scrub where it occurs today on Mounts Jaya and Trikora in the Snow Mountains. ●

AGE
Late Pleistocene, estimated to be 25 000–20 000 years ago.

LOCALITY
Kelangurr Cave in the Baliem River catchment near Kwiyawagi, central montane Irian Jaya, Indonesia.

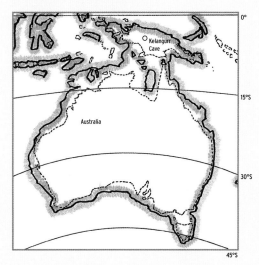

ENVIRONMENT
The fauna suggests that these animals lived in an alpine tussock grassland and scrub habitat. Today the area around the cave is tall upper montane forest. Hence, this cave predates the loss of grasslands from this area, a transition almost certainly driven by climate change.

REFERENCES
116, 223, 236, 271

Nombe Rock Shelter, Papua New Guinea
PLEISTOCENE

EXCAVATIONS OF NOMBE ROCK SHELTER IN NEW Guinea since 1964 have revealed a complex, often human-disturbed late Pleistocene record. However, the general pattern of what happened is reasonably clear. During the end of the last glacial maximum, around 25 000 years ago, there were now extinct megafaunal species such as *Protemnodon tumbuna* present in the Highlands. Most of them disappeared around 14 000 years ago, the time when glaciers began to retreat, and the highland forests gave way to alpine grasslands. Although humans were present probably as early as 60 000 years ago, there is no evidence that they hunted the extinct megafaunal species, which probably declined in response to late Pleistocene climate change in the Highlands. Nevertheless, the extinct megafaunal kangaroos at Nombe survived perhaps 15 000 years longer than their counterparts in Australia. Species of *Protemnodon* at Nombe (far right) appear not to be closely related to contemporary Australian species of this genus; rather, they were descendants of immigrants that migrated from Australia to New Guinea during the Pliocene. The

Thylacine *Thylacinus cynocephalus* was present in the Nombe deposits for the entire history from more than 24 000 years ago to at least 5000 years ago, indicating that Thylacines disappeared from New Guinea about the same time they did on the Australian mainland. These disappearances may have been caused by the introduction of dogs about the same time to both areas. The living White-bibbed Fruit Dove *Ptilinopus rivoli* inhabits, among other environments in New Guinea, montane forests potentially of the kind that existed around Nombe 25 000 years ago. ●

AGE

Late Pleistocene, about 25 000-5000 years ago.

LOCALITY

Nombe Rock Shelter on the eastern slopes of the Highland Erimbari limestone escarpment, Simbu Province, New Guinea.

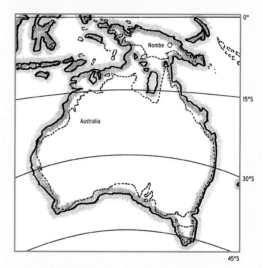

ENVIRONMENT

In late Pleistocene times, before it was naturally eroded to become a long shelter with a small cave at the rear, Nombe was a large cave with a river flowing through it. When the extinct species were alive, Nombe was close to the treeline. Smaller animals in the deposit suggest Thylacines and/or humans brought in prey from adjacent grasslands as well as forests, but the environment around the cave itself when the extinct species were present may have been alpine rainforest.

REFERENCES

121, 223, 264, 375

South Island moa, New Zealand
PLEISTOCENE/HOLOCENE

OF THE SIX GENERA AND NINE SPECIES OF extinct moa, the most bizarre-looking has to be the Heavy-footed Moa *Pachyornis elephantopus* (shown here with two young). Just a little over 1 m high at its back and about 75 cm wide, reaching over 200 kg in weight and with its tummy nearly dragging on the ground, this moa was the most extreme of all graviportal birds known. Its name, given by Sir Richard Owen, reflects the pachydermal proportions of its tarsus. It was restricted to the lowland shrublands in the South Island and was a major component of Canterbury and eastern Otago faunas during the Holocene. It was absent in the western regions except during glacial intervals of the Pleistocene when shrublands became available in these areas. It is one of three species in the genus. The Crested Moa *P. australis* lived in the shrublands of the subalpine zone and the smaller Mappin's Moa *P. geranoides* lived in the North Island. The Heavy-footed Moa had a very stout and short bill that effectively acted like secateurs, enabling it to eat extremely fibrous plants such as flax *Phormium tenax*, leaves of which have been found in preserved gizzards associated with its skeletons. In the subcanopy is another of New Zealand's extinct endemic birds, the South Island Piopio *Turnagra capensis*. This is actually an oriole, long separated from its relatives in Australia. It had a sister species in the North Island. Both are among the ancient passerine groups of New Zealand, along with the wattle birds (Huia, kokako and saddlebacks), New Zealand wrens and *Mohoua* species. Piopio species were omnivores, gleaning insects from leaves, eating invertebrates on the forest floor and gobbling berries when available. ●

AGE

Late Pleistocene to Holocene, about 100 000–750 years ago.

LOCALITY

South Island, New Zealand.

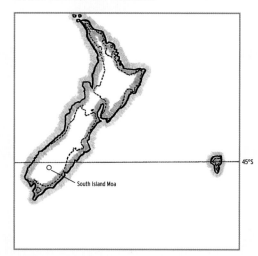

South Island Moa

45°S

ENVIRONMENT

Lowland eastern forest with grassy glades. Forest species included giant red beech trees *Nothofagus fusca* and flax plants *P. tenax* which had strap-like leaves. A few mosses covered the forest floor. Seasonal droughts were common.

REFERENCES

322, 452, 471

South Island adzebill, New Zealand

ONE OF THE MOST BIZARRE EXTINCT BIRDS THAT once inhabited the New Zealand landscape was the adzebill, with a species in the North and South Islands. Adzebills, classified in their own family Aptornithidae, were large birds, standing about 75 cm tall and weighing in at about 15 kg, with a fearsomely large bill. These birds are distantly related to rails and are in the order Gruiformes. Surprisingly, DNA evidence has conclusively shown that their nearest relatives are the tiny sarothrurid rails now restricted to Africa and Madagascar. Like many gruiforms, these birds were probably omnivores although isotopic analyses of their bones suggest that they preyed on vertebrates. New Zealand had lots of frogs and lizards as well as birds that were potential prey. In the scene portrayed here, the South Island adzebill *Aptornis defossor* has captured a large gecko *Hoplodactylus duvaucelii* which, growing to around 30 cm, is the largest lizard in New Zealand. It still survives in a few protected areas, mainly on islands that are rodent-free. Adzebills were entirely flightless: their wings were reduced to tiny vestigial structures, relatively little larger than the wings of kiwis. They lived in New Zealand for a long time. An ancestral form is known from the early Miocene fossil deposits near St Bathans in Otago, New Zealand. During the Holocene, the South Island adzebill was restricted to vegetation mosaics of the eastern part of the South Island. During the glacial periods, its range was far larger, including areas in the then deforested western regions. However, not long after these birds encountered humans around 750 years ago, they were extinct. This loss meant that Western naturalists never had a chance to study one of the most enigmatic components of New Zealand's avifauna. ●

AGE

Late Pleistocene to late Holocene, about 100 000–750 years ago.

LOCALITY

Foothills of Canterbury Plains, Geraldine, South Canterbury, New Zealand.

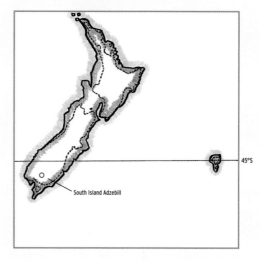

South Island Adzebill

45°S

ENVIRONMENT

In a small forest grove on one of the low limestone hills, giant matai (*Prumnopitys taxifolia*) trees grew over 35 m tall, shading the mosses that covered the forest floor. Not far away, through a bank of shrubs, there were more open habitats dissected by gravel streams with grass-covered margins.

REFERENCES

56, 283, 384, 411, 452

THE QUIET OF A MORNING IN THE BROADLEAF
forests on the hillside are suddenly interrupted
by an urgent rustle of the forest litter. A rush of
movement heralds the charge out of the bushes
of a 3 m crocodilian *Volia athollandersoni*, and
with a squawk the unsuspecting Noble Megapode
Megavitiornis altirostris tries to flee. *Volia
athollandersoni*, a mekosuchine crocodile, was
named after the place where this scenario played
out (Volivoli Cave) and the eminent archaeologist
Professor Atholl Anderson. It was undoubtedly the
top predator in this land. Like other mekosuchine
crocodiles, this one appears to have been more
terrestrial in its lifestyle, straying far from the
watery habitats favoured by most living crocodiles.
It was trapped in Volivoli Cave which is up a hill,
several hundred metres from the closest wetland
below. It had diverse animals to prey on, with large
frogs, iguanas, turtles, snakes and flightless birds
including a dodo-sized pigeon and megapodes.
The Noble Megapode was 75 cm tall and weighed
perhaps 15 kg. It, and a giant flightless pigeon (a
species of *Natunaornis*), were the equal largest
birds in Fiji. Like its relative *Sylviornis neocaledoniae*
in New Caledonia, this giant flightless megapode
had a huge deep bill and a large head. It has been
recorded in the earliest human sites in Fiji, dated
at about 3000 years old, It may well have been an
early casualty of human arrival in this archipelago.

●

AGE
Late Pleistocene-Holocene, about 20 000–
3000 years ago.

LOCALITY
Volivoli Cave, Sigatoka River, Vitilevu, Fiji.

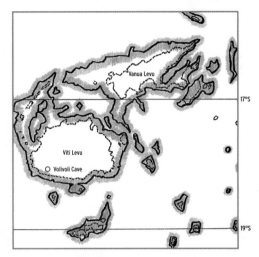

ENVIRONMENT
The broad Sigatoka River winds down the
valley towards the south, draining a large
part of Vitilevu's seasonally dry western
region. The hills flanking the western side, a
few kilometres in from the sea, are limestone
and clothed in a diverse broadleaf forest. At
the back of a rock shelter facing east, a dark
pit drops into a void below. Here multiple
animals met their death, having ventured one
step too far back while sunbaking or seeking
shelter from the monsoonal torrential rains.
Never mind the reason, that pit, as part of
the Volivoli Cave, became a rich record of
intriguing animals now lost from the island
of Fiji.

REFERENCES
185, 261, 443, 444, 447

Tongoleleka, Tonga

HOLOCENE

THE EXTINCT IGUANA *BRACHYLOPHUS GIBBONSI*
sunbathes in a forest glade, while two large and
brightly coloured pigeons stay in the shadows.
This iguana, reaching 1.2 m long, was the largest
species in its genus and the Tongan representative
of a genus otherwise found then and now only
in Fiji. They are assumed to have colonised these
islands following long-distance dispersal from the
Americas, far to the east. The pigeon, a species of
Caloenas similar in size to the extinct *C. canacorum*
of New Caledonia, was only one of many species of
pigeons in the Tongan forests, but while it was not
the largest – there were two larger species (*Ducula
shutleri* and *Tongoenas burleyi*) – it probably had
the most spectacular plumage. The sole surviving
member of this genus is the Nicobar Pigeon
Caloenas nicobarica found far to the west in a
region stretching from the Solomons to Nicobar
Islands. Extinction of large taxa like these has been
a common outcome following human arrival on
many of the Pacific islands. However, inadvertent
introduction at the same time of the Pacific Rat
Rattus exulans, which is known to predate nesting
birds and eggs, into this and many other Pacific

islands may have been a significant cause of some of the extinctions that followed. Current efforts to eradicate this introduced rat from many of these islands, including those of Tonga, are based on research which shows they continue to constitute a major threat to existing birds on these islands. In the Ha'apai Group, where the birds depicted here once lived, Lapita voyagers from Fiji arrived about 2800 years ago, probably with Pacific Rats aboard. Whenever these large birds and several other extinct birds were eaten by the Lapita people, the remains ended up in the middens of the Tongoleleka Site on Lifika. Over half of the original bird species in Tonga and many other Pacific islands did not survive very long after the arrival of humans and Pacific Rats, but fortunately these enduring Polynesian middens provide a window back into time to document the species that were lost. ●

AGE
Holocene, about 10 000–2800 years ago.

LOCALITY
Lifika, Ha'apai Group, Tonga.

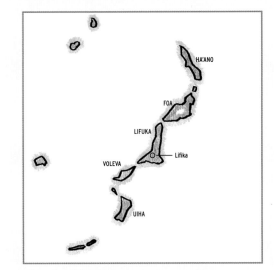

ENVIRONMENT
On the western side of a small island fringed in coconut palms, white sandy beaches are protected from the fury of the ocean by an enclosing coral reef. Passing under the coconut trees and around the pandanus groves, one quickly enters the broadleaf forest where regular rain ensures the ferns never dry out too much.

REFERENCES
309, 310, 365, 366, 367, 446

Chatham Islands, New Zealand

THE QUIET WATERS PROVIDED ABUNDANT FOOD
for a wide diversity of birds. Seals, such as the
New Zealand Sea Lion *Phocarctos hookeri* and the
Fur Seal *Arctocephalus forsteri* (centre right), also
visited this lagoon, which is periodically open to
the sea, to rest from stormy conditions at sea.
Swimming on the lagoon, huge flocks of Black
Swans *Cygnus atratus* browsed the vast beds of
macrophytes under its surface. The swan, also
found on mainland New Zealand 800 km to the
west, was hunted to extinction by the first visitors
before being reintroduced in the 19th century by

Europeans to areas of mainland New Zealand, from
whence it recolonised the Chatham Islands. It is
thus a rare example of a native species that has
benefited from the pastoralisation of New Zealand.
Along with the swans lived many species of ducks.
Perhaps the first duck to reach the Chathams
was the Chatham Island Duck, formerly regarded
as a distinct taxon (*Pachyanas chathamica*) on
account of its large size, but now regarded to be
an overgrown teal *Anas chathamica*. This larger
size of an island form runs counter to the trend
more often seen in insular evolution of ducks,

where the island forms are smaller. Along the shore, the Chatham Island Coot *Fulica chathamensis* searches for invertebrates. Nearby, an inquisitive parrot – the recently described Chatham Island Kaka *Nestor chathamensis* – cheekily investigates a resting seal. The kaka, coot and duck are all closely related to living forms on mainland New Zealand, their differences resulting from only 1–2 million years of evolution in isolation because the islands themselves are no older than this. ●

AGE
Holocene, about 10 000–600 years ago.

LOCALITY
Te Whanga Lagoon, Chatham Island, New Zealand.

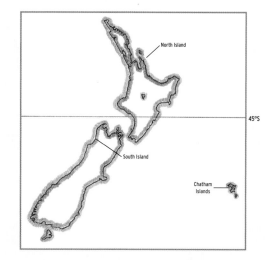

ENVIRONMENT
This was a large (about 15 × 5 km) shallow lagoon occupying most of the centre of Chatham Island. Regularly breached by the sea, the waters were usually somewhat salty. Large beds of sea grass grew in its shallows. To the west and south, low peat-cloaked hills surrounded the lagoon. Low forests covered the hills to the north and south of this area.

REFERENCES
251, 252, 254

North Island, Scene 1, New Zealand

PRIOR TO THE ARRIVAL OF HUMANS ABOUT
750 years ago, New Zealand was a land of birds.
In the North Island, dominating this avifauna in all
senses of the word were several species of moa
(Dinornithiformes). Largest of all was the North
Island Giant Moa *Dinornis novaezealandiae* (centre
left), with the larger females able to reach 3 m high
into the trees to browse. These giants may have
weighed up to 200 kg. They lived in all habitats
from the coastal dunes to the deep forests, but
other smaller species, such as Mantell's or Mappin's
Moa *Pachyornis geranoides* (background near
waterfall) and Coastal Moa *Euryapteryx curtus
curtus* (far right), were restricted to mosaics of
grassland and shrubland. The dense forest was
the home of the Little Bush Moa *Anomalopteryx
didiformis*, where it browsed alongside small
foraging rails such as the Snipe Rail *Capellirallus*

karamu (lower far left), seen here at the water's
edge, while the Huia *Heteralocha acutirostris* gave
forth melodic calls the branch overhead. Huia are
famous for the fact that the sexes have different-
shaped bills – that of the female is longer and more
curved – and for their white-tipped tail feathers
that are highly valued by Māori and Europeans
alike. Along the forest margin, two large North
Island Geese *Cnemiornis gracilis* with chicks grazed
on grasses and other herbs. It was related to the
Cape Barren Goose of Australia, but its lineage
has been in New Zealand since at least the early
Miocene about 19–16 million years ago. Many
waterfowl occurred on pools developed in larger
streams. Here two Scarlett's Duck *Malacorhynchus
scarletti* alight on the pool. These are large
relatives of the unique Australian Pink-eared Duck
M. membranaceus and were specialist dabbling

ducks. The odd duck shown with a large pouch under its bill is the New Zealand Musk Duck *Biziura delautouri*, another endemic New Zealand species with a living relative in Australia. All of the species shown here went extinct soon after humans arrived. Factors involved in these losses probably included forest clearance and predation by both the people and the Pacific Rat *Rattus exulans* that the first arriving humans introduced to these islands. The Huia survived till European times, but European rats, stoats and hunting for its tail feathers soon led to its demise as well. ●

AGE
Holocene, about 10 000–600 years ago.

LOCALITY
North Island, New Zealand.

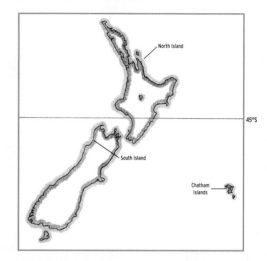

ENVIRONMENT
During the Holocene, the North Island was mostly clothed in ever-wet dense podocarp-hardwood forests with a dense interlocking canopy that shielded abundant mosses and ferns on the forest floor. Only on some dune areas, or along river margins and in areas recently deforested by volcanic activity, did more open habitats exist for those species that preferred shrublands and grasslands.

REFERENCES
384, 452, 456

North Island, Scene 2, New Zealand

MANY NOW-EXTINCT FLIGHTLESS BIRDS WERE
among the most conspicuous components in
the environments of the North Island of New
Zealand. They included one of the largest birds
ever to exist, the North Island Giant Moa *Dinornis
novaezealandiae* and some of the smallest, such
as diminutive wrens. Here, the sparrow-sized North
Island Stout-legged Wren *Pachyplichas jagmi*
is dwarfed by a passing moa. The stout-legged
wrens – for there is a South Island species as well,
along with the Long-billed Wren *Dendroscansor
decurvirostris* and Lyall's Wren *Traversia lyalli* – are
famous as four of the only six known flightless
passerines. Passerines, or songbirds, dominate the
world's avifauna, comprising some 5300 of the
approximately 10 000 known bird species, yet only
on the Canary Islands in the Atlantic, where there
are two species of flightless greenfinches, are there
any other flightless ones. But the New Zealand
wrens are famous for something else in addition
to their weak or absent flying abilities – this group
of five genera and seven species, members of
the family Acanthisittidae, constitutes the sister
group to all other songbirds on Earth. Today, New
Zealand's smallest bird, the Rifleman *Acanthisitta
chloris*, and the Rock Wren *Xenicus gilviventris*
remain as the only living members of this family
which is important for our understanding about
the origins of all other passerines. It is possible
that these tiny weak-flying birds have existed on
Zealandia ever since that continental fragment
separated from Gondwana 80–60 million
years ago. However, to date the oldest known
acanthisittid wren *Kuiornis indicator* comes from
the 19–16 million years old St Bathans fauna from
central Otago, New Zealand, and the oldest-
known passeriform bird comes from the 55 million
years old Tingamarra Local Fauna from Murgon,
Queensland (p. 96). ●

AGE

Late Pleistocene to Holocene, about
100 000–750 years ago.

LOCALITY

North Island, New Zealand.

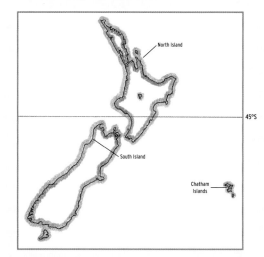

ENVIRONMENT

In the closed canopy forests of the North
Island, where sunlight never reached the
forest floor, mosses and litter covered the
ground and provided habitat for abundant
invertebrates. Lacking rodents before
humans arrived, New Zealand evolved faunas
rich in large invertebrates and a multitude
of ground-dwelling flightless birds to prey on
them.

REFERENCES

178, 250, 451

New Caledonia's giant megapode

ONE OF THE MOST CONSPICUOUS ANIMALS IN
the prehistoric New Caledonian landscape was
undoubtedly *Sylviornis neocaledoniae*, a giant
megapode. Megapodes are unique in the avian
world for being obligate ectothermal incubators –
that is, they use external heat sources to incubate
their eggs. Some dig holes and lay their eggs in
warm soils in active volcanic landscapes, but most
construct large mounds of soil and decomposing
leaf litter in which the eggs are laid. The birds
monitor and control the temperature in the
mounds by variously increasing and decreasing
the thickness of covering soil. They literally dig and
scratch all day. This New Caledonian megapode
was the largest of the group. It was about 1 m
tall and probably weighed around 20 kg. It was
very different from other megapodes and for
this reason has been placed in its own family, the
Sylviornithidae. It had markedly reduced wings
and certainly could not fly. Moreover, its skull
and especially its beak had evolved a very un-
megapode-like form. The bill was deep with a large
bony casque that extended back over the skull.
This amazing structure is not, however, novel. A
distantly related galliform, the Helmeted Curassow
Pauxi pauxi of South America, sports a similar
embellishment. As in this Curassow, the helmet
was probably coloured and used for behavioural
displays. Unfortunately, *S. neocaledoniae*
disappeared shortly after humans arrived on New
Caledonia about 3000 years ago, without leaving
any clues about its natural history including what it
probably ate. ●

AGE

Holocene, about 10 000–2800 years ago.

LOCALITY

Nepouri Peninsula, Grande Terre,
New Caledonia.

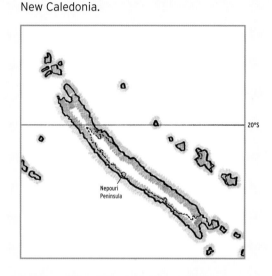

ENVIRONMENT

When these species were alive, there were, as
there are today, rainforests along the eastern
side and dry sclerophyll forests on the
western side of the peninsula. These lowland
forests provided abundant litter and soil to
scratch into giant mounds, and invertebrates
as prey.

REFERENCES

32, 265, 304

Canterbury Plains, New Zealand
HOLOCENE

ON THE LOWLANDS OF THE CANTERBURY PLAINS, a diverse range of habitats provided homes to the richest avifauna existing in New Zealand during the Holocene. Portrayed here, greeting each other, are a pair of the heaviest and tallest birds ever to exist in New Zealand, the South Island Giant Moa *Dinornis robustus*. Moas belonging to the genus *Dinornis* had the greatest degree of sexual size dimorphism known in birds. The females were more than twice the size of the males. Adding further to the variation in this species, we now know that their size considerably varied across the landscape. Only 100 km inland, in the hill country, mean individual size was markedly smaller and, going up-slope into the montane regions, size was further reduced such that in the high beech forests the larger female was the same size as the smaller males of lowland Canterbury. Size variation of this kind originally led to overestimation of the number of species. This error was resolved only with the advent in recent years of ancient DNA studies. Other moas that co-existed with the species of *Dinornis* in the Canterbury forests included the specialist lowland Eastern Moa *Emeus crassus* and the South Island morph of the Coastal Moa *Euryapteryx curtus gravis* (two individuals in background). The latter was considerably larger than its North Island relative and weighed in at around 150 kg. All of these extinct moa feature prominently in Māori middens that were accumulated in the 100 years after humans first arrived in New Zealand 750 years ago, accompanied by Pacific Rats *Rattus exulans*. In the canopy, the Orange-wattled Crow or South Island

Kokako *Callaeas cinerea* fed on fruit and leaves. Soaring above it over the forest gaps was a giant relative of the Swamp Harrier, the extinct New Zealand Eyles' Harrier *Circus teauteensis*, on the lookout for a careless kokako or fat pigeon. ●

AGE
Holocene, about 10 000–750 years ago.

LOCALITY
Canterbury Plains, South Island, New Zealand.

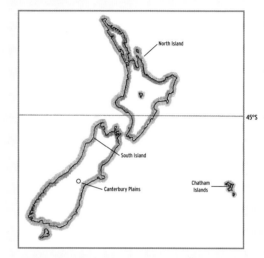

ENVIRONMENT
In the lowlands of the Canterbury Plains, over thousands of years meandering rivers had created a mosaic of swamp forests, shrublands and grasslands. Tall podocarp trees such as Matai *Prumnopitys taxifolia* provided abundant seasonal resources of fruit and towered over understorey trees. Notable in the shrub layer were large numbers of spiny divaricating plants and other forms adapted to resist the browse of moas. These included the Lancewood *Pseudopanex crassifolius*, which had tough elongate juvenile leaves as shown near one of the moas.

REFERENCES
150, 179, 442

New Zealand giant eagle

A LARGE FEMALE HAAST'S EAGLE *HIERAAETUS moorei*, formerly *Harpagornis moorei*, attacks a Māori man. This eagle likely was resting on a high vantage spot, either on a nearby crag or the top of a large tree, when it saw the man. Perhaps hungry in the absence of its once favoured moa prey, it risked an attack on this novel animal, possibly not understanding the risk it was taking in doing so. Reaching 13 kg in weight with a wingspan of about 3 m, this eagle was the largest in the world. It was restricted to the South Island of New Zealand where forest margins and more open habitat facilitated its hunting. There is abundant evidence of it preying on moa, with bones of Eastern

Moa, Coastal Moa and even Giant Moa showing distinctive damage. Typically, this eagle landed on the back of the moa, one 30 cm wide set of claws straddling the pelvic region and the other further up the moa's trunk. The giant claws of the eagle's hind toe were capable of slicing through the feathers and skin, 5–10 cm of muscle, then the 5 mm thick bony plates of the pelvis, leaving great gashes in this bone upwards of 10 cm long. It is probable that the kidneys were then ripped apart, which would have triggered the moas to die soon after of shock. Such is the story recorded in the many moa pelvises that have been found in swamp deposits. Humans – of much smaller stature than some moas and less protected by a pelage – would certainly have been potential prey, given that the much smaller golden eagles of Eurasia can and do kill people. Haast's Eagle had a smaller ancestor, the Little Eagle *Hieraaetus morphnoides* of Australia, which is only a 10th the size its New Zealand descendant. Surprisingly, this ancestor appears to have colonised New Zealand perhaps only a million years ago, if the DNA evidence is to be believed, after which it must have rapidly increased in body size. Support for this origin hypothesis may well be the relatively small size of the brain of Haast's Eagle. While the occasional Māori may have met their end through struggles with this huge predator, it soon joined the moas on the long register of extinctions correlated with the arrival of Māori in New Zealand. Bones of this giant eagle are commonly found in the middens of the first Māori villages. ●

AGE
About 700 years ago.

LOCALITY
South Island, New Zealand.

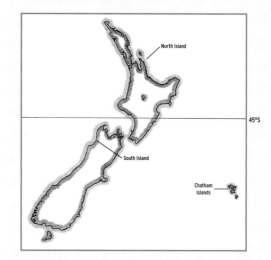

ENVIRONMENT
The scenario pictured here might have taken place in one of the alpine valleys of the South Island or along the margins of one of the great braided rivers near the forest edge.

REFERENCES
65, 67, 152, 350, 452

Pindai Cave, New Caledonia

HOLOCENE

IN AN ANCIENT FOREST RAUCOUS WITH LIFE, A crystal-clear stream trickles down a valley. Here, a small semi-terrestrial deep-headed crocodile *Mekosuchus inexpectatus* is cracking the shell of a *Placostylus duplex* snail it found on rocks along the edge of the stream. A giant terrestrial meiolaniid turtle waits patiently to come down to drink, while nearby a large flightless swamphen *Porphyrio kukwiedei* nervously watches the crocodile's every move. As its name records, it was a great surprise to scientists when they discovered this strange fossil crocodile on the island of Grande Terre, New

Caledonia. It is now known to have been one of the last surviving members of the distinct subfamily Mekosuchinae, whose members were also known from Australia, Vanuatu, Fiji and probably New Zealand. The large horned meiolaniid turtle shares a similar distribution pattern to species of *Mekosuchus* and likewise is entirely extinct. The swamphen is a rail of the family Rallidae, a group notorious for their ability to disperse widely across water gaps, but also paradoxically for their tendency to evolve into subsequently flightless species on far-flung islands. This one belonged to the purple swamphen lineage, a group represented by flying forms that occur in lands stretching from Eastern Europe to Australia and far out into the Pacific. On New Caledonia they became huge, approaching the size of the more famous and extant New Zealand relative the Takahe *P. hochstetteri*. They thrived on Nepouri Peninsula, a small area of limestone that houses the Pindai Caves, some of which made excellent pitfall traps. All these New Caledonian animals, along with another 45 species of birds (20 of which are now extinct), had their bones preserved in these caves. The turtles, crocodiles and this swamphen became extinct shortly after humans arrived about 2800 years ago. ●

AGE

Holocene, about 6000-2800 years ago.

LOCALITY

Nepouri Peninsula, Grande Terre, New Caledonia.

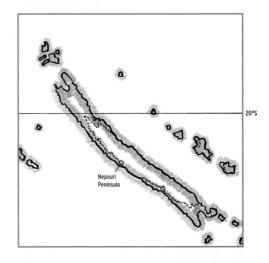

ENVIRONMENT

Today and probably when these fossils were accumulated, the western coast of Grande Terre supports a dry sclerophyll forest on the lowlands near the coast. But around Pindai Caves on Nepouri Peninsula, the roots of broadleaved tropical trees penetrate the porous limestone, enabling this lusher vegetation to flourish.

REFERENCES

7, 30, 31, 32

South Island goose, New Zealand

ONE OF THE MOST SPECTACULAR AVIAN

inhabitants to be met by Māori upon their arrival in New Zealand were species of geese in the genus *Cnemiornis*. The largest was *C. calcitrans* from the South Island, weighing in at about 15 kg and standing over 1 m tall. These geese lived only in lowland open habitats. They were absent from closed forest environments. During the Holocene, *C. calcitrans* was restricted to eastern parts of the South Island where it grazed upon grasses and herbs, probably in the same way its nearest relative, the Cape Barren Goose *Cereopsis novaehollandiae* of Australia does today. This Australian goose creates close-cropped swards of the plants on which it feeds, and it is probable that *Cnemiornis*, with its squared-off bill, would have created similar mown lawns in their favoured breeding areas. The *C. calcitrans* lineage has been a resilient one in New Zealand, with earlier members whose bones have been found in the St Bathans fossil deposits, suggesting that it has been around since at least the early Miocene. Following the early Miocene, species of *Cnemiornis* became greatly modified from their Cape Barren Goose-like ancestor, including complete loss of their ability to fly. This trait may well have contributed to the downfall of the species after the arrival of Māori, within perhaps 100 years. They shared the open habitats with another endemic species, the Black Stilt *Himantopus novaezelandiae*. The latter breeds only in the braided riverbeds of the South Island, secreting its nest and eggs among the river stones. Although it still survives, it is highly endangered because of nest predation by introduced mammals such as the omnivorous Pacific Rat *Rattus exulans* that arrived with the Māori, loss of habitat and competition from the more recent colonist the Pied Stilt *H. leucocephalus*. ●

AGE

Late Holocene, about 1000 years ago.

LOCALITY

Canterbury Plains, South Island, New Zealand.

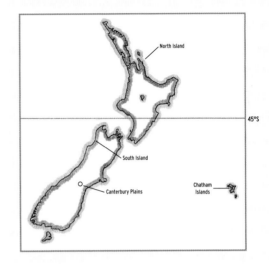

ENVIRONMENT

This was a scene typical of the edge of a river in Canterbury, where gravel-lined flood channels wound between tussock grasses. The open gravel beds provided nesting habitat for stilts and other riverine birds, such as dotterels, while the tussocks and intervening herbs provided browse for grazing geese.

REFERENCES

282, 452, 453

Acknowledgements

Mike Archer. On behalf of Sue and myself, we thank many professional colleagues who have been instrumental in helping us to reveal the fascinating deep-time history of Australia and its surrounding lands. I am very grateful in particular to my UNSW research colleagues Suzanne Hand, Karen Black, Henk Godthelp, Phil Creaser, Troy Myers, Anna Gillespie, Georgina Hickey, Stephan Williams, Ian Graham and Mike Augee. Our research at Riversleigh and Murgon has been facilitated by the Elders of the Waanyi Nation (the traditional owners of Riversleigh), Riversleigh/Lawn Hill Station, the Australian Research Council, Queensland, and Australian Museums, Queensland Parks and Wildlife Service, Environment Australia, IUCN World Heritage Committee, UNSW Sydney, Riversleigh World Heritage Management Committee, Riversleigh Interpretive Centre, Arthur and Karen White and Riversleigh Society Inc., National Geographic Society, Mount Isa Mines, Adel's Grove, Phil Creaser's CREATE Fund, Ken and Margaret Pettit, Doug and Anne Jeanes, Elaine Clarke, Alan Bartholomai, Martin Dickson, Tim Cody, Margaret Beavis, Peter Bridge, Dick Smith, the Australian Geographic Society, Gary Johnston, Tony McGrady, Ron McCullough, the family of Alan Rackham, Michael Wentworth, John Prince, James Porter, Chris Braithwaite, Chris McInerny, Leo and Glenda Geraghty and so many others. Thanks also to my skilled co-authors on this project, Peter Schouten, Suzanne Hand, Trevor Worthy, John Long, as well as the highly professional team at CSIRO Publishing.

Suzanne Hand. On behalf of Mike and myself, a huge thankyou to the vast and growing army of extraordinary volunteers, exceptional students and brilliant researchers from across the globe who, since 1978, have given years of their lives to help interpret the fossil record at Riversleigh and Murgon. Colleagues who helped bring our research at Riversleigh to the world's attention include Sir David Attenborough, Bob Beale, Quentin Jones, Geoff Burchfield, Louise Egerton, Ross Arnett, Dorothy Dunphy, Alan Pryke, Robyn Williams, Nicky Phillips, Tony Walters, Leigh Dayton and James Woodford, to name just a few. I would also like to thank my co-authors Mike Archer, John Long and Trevor Worthy for the wonderful opportunities they have given me over many years, and to Peter Schouten for breathing new life into a hoard of our favourite fossils.

Trevor Worthy. I thank many people whose inspiration helped to reveal the fascinating world that this book examines. Principle among these were Sir Charles Fleming, John Yaldwyn, Storrs Olson, Walter Boles and David Steadman. I am grateful to numerous people that I have collaborated with, especially from New Zealand – J.A. (Sandy) Bartle, Richard Holdaway, Vanesa De Pietri, Paul Scofield, Alan Tennyson; Australia – Mike Archer, Suzanne Hand, and Aaron Camens; and the Pacific – Atholl Anderson, Christophe Sand. I thank all those landowners who enabled the exploration of fossil deposits on their land, especially Ann and Euan Johnstone, St Bathans, NZ. I also wish to thank the museums for the vital role they have in preserving specimens that underpin our research. In addition to those mentioned by John Long and Mike Archer, the Auckland Museum, Waitomo Caves Museum, the Museum of New Zealand, Canterbury Museum and Otago Museum maintain key collections that hold the fossils of the prehistoric animals herein portrayed. Importantly, I am extremely grateful to my wife Jenny Worthy who has actively indulged in and joined my passion for palaeontology to help reveal the amazing world of fossils.

John Long. I acknowledge the Gooniyandi people of the Fitzroy Crossing area on whose land we have worked over the past 38 years, and the support for research and access to collections from the Western Australian Museum, Museum Victoria, the Australian Museum, South Australian Museum and Queensland Museum. The Australian Research Council and the National Geographic Society funded many research trips. Finally, I thank many colleagues for helpful discussions over the years including Pat Vickers-Rich, Tom Rich, Gavin Young, Alex Ritchie, Sue Turner, Carol Burrow, Kate Trinajstic, Alice Clement, Tim Flannery, Steve Salisbury, Ralph Molnar, Scott Hocknull, Anne Warren, Mike Lee, Mike Archer and Suzanne Hand. My work would be impossible without the hard work of this legion of truly dedicated people.

Peter Schouten. My art would be impossible without the hard work of a legion of dedicated fossil discoverers, preparators and all the time-travelling palaeontologists who have described their finds. There are in fact too many to name, so thank you to them all! Most importantly I would like to thank my co-authors Sue Hand, Michael Archer, John Long and Trevor Worthy, not just for their contribution to this book, but also for the many decades of support, guidance and belief in my abilities. Thanks also to Briana Melideo of CSIRO Publishing for her belief in this project.

References

1. Agnolin FL, Ezcurra MD, Pais DF, Salisbury SW (2010) A reappraisal of the Cretaceous non-avian dinosaur faunas from Australia and New Zealand: evidence for their Gondwanan affinities. *Journal of Systematic Palaeontology* **8**, 257–300. doi:10.1080/14772011003594870

2. Aguirre-Fernández G, Fordyce RE (2014) *Papahu taitapu*, gen. et sp. nov., an Early Miocene stem Odontocete (Cetacea) from New Zealand. *Journal of Vertebrate Paleontology* **34**, 195–210. doi:10.1080/02724634.2013.799069

3. Ahlberg PE, Johanson Z (1997) Second tristichopterid (Sarcopterygii, Osteolepiformes) from the Upper Devonian of Canowindra, New South Wales, Australia, and phylogeny of the Tristichopteridae. *Journal of Vertebrate Paleontology* **17**, 653–673. doi:10.1080/02724634.1997.10011015

4. Ahlberg PE, Johanson Z, Daeschler EB (2001) The Late Devonian lungfish *Soederberghia* (Sarcopterygii: Dipnoi) from Australia and North America, and its biogeographic implications. *Journal of Vertebrate Paleontology* **21**, 1–12. doi:10.1671/0272-4634(2001)021[0001:TLDLSS]2.0.CO;2

5. Akerman K (1973) Two Aboriginal charms incorporating fossil giant marsupial teeth. *Western Australian Naturalist (Perth)* **12**, 139–141.

6. Allwood AC, Walter MR, Burch IW, Kamber BS (2007) 3.43 billion-year-old stromatolite reef from the Pilbara Craton of Western Australia: ecosystem-scale insights to early life on Earth. *Precambrian Research* **158**, 198–227. doi:10.1016/j.precamres.2007.04.013

7. Anderson A, Sand C, Petchey F, Worthy TH (2010) Faunal extinction and human habitation in New Caledonia: initial results and implications of new research at the Pindai Caves. *Journal of Pacific Archaeology* **1**, 89–109.

8. Andrews SM, Long JA, Ahlberg P, Barwick RE, Campbell KSW (2006) The structure of the sarcopterygian *Onychodus jandemarrai* n.sp. from Gogo, Western Australia: with a functional interpretation of the skeleton. *Transactions of the Royal Society of Edinburgh. Earth Sciences* **96**, 197–307. doi:10.1017/S0263593300001309

9. Aplin KA, Rich TH (1985) *Wynyardia bassiana* Spencer, 1901. The Wynward marsupial. In *Kadimaka: Extinct Vertebrates of Australia.* (Eds PV Rich and GF van Tets) pp. 219–224. Princeton University Press, Princeton.

10. Archer M (1977) Origins and subfamilial relationships of *Diprotodon* (Diprotodontidae, Marsupialia). *Memoirs of the Queensland Museum* **18**, 37–39.

11. Archer M (1978) Quaternary vertebrate faunas from the Texas Caves of southeastern Queensland. *Memoirs of the Queensland Museum* **19**, 61–109.

12. Archer M (1978) Australia's oldest bat, a possible rhinolophid. *Proceedings of the Royal Society of Queensland* **89**, 23.

13. Archer M (1982) A review of the dasyurid (Marsupialia) fossil record, integration of data bearing on phylogenetic interpretation, and suprageneric classification. In *Carnivorous Marsupials.* (Ed. M Archer) pp. 397–443. Royal Zoological Society of New South Wales, Sydney.

14. Archer M, Arena DA, Bassarova M, Beck R, Black K, Boles WE, Brewer P, Cooke BN, Crosby K, Gillespie A, Godthelp H, Hand SJ, Kear B, Louys J, Morrell A, Muirhead J, Roberts KK, Scanlon JD, Travouillon KJ, Wroe S (2006) Current status of species-level representation in faunas from selected fossil localities in the Riversleigh World Heritage Area, northwestern Queensland. *Alcheringa* **30**(supp.1), 1–17. doi:10.1080/03115510609506851

15. Archer M, Bates H, Hand SJ, Evans T, Broome L, McAllan B, Geiser F, Jackson S, Myers T, Gillespie A, Palmer C, Hawke T, Horn AM (2019) The Burramys Project: a conservationist's reach should exceed history's grasp, or what's the fossil record for? *Philosophical Transactions B* 374, 20190221.

16. Archer M, Binfield P, Hand SJ, Black KH, Creaser P, Myers TJ, Gillespie AK, Arena DA, Scanlon J, Pledge N, Thurmer J (2018) *Miminipossum notioplanetes*, a Miocene forest-dwelling phalangeridan (Marsupialia; Diprotodontia) from northern and central Australia. *Palaeontologia Electronica* **21**, 1–11. doi:10.26879/757

17. Archer M, Crawford IM, Merrilees D (1980) Incisions, breakages and charring, some probably man-made, in fossil bones from Mammoth Cave, Western Australia. *Alcheringa* **4**, 115–131. doi:10.1080/03115518008619643

18. Archer M, Flannery TF (1985) Revision of the extinct gigantic rat kangaroos (Potoroidae: Marsupialia) with a description of a new Miocene genus and species and a new Pleistocene species of *Propleopus*. *Journal of Paleontology* **59**, 1331–1349.

19. Archer M, Flannery TF, Ritchie A, Molnar RE (1985) First Mesozoic mammal from Australia: an early Cretaceous monotreme. *Nature* **318**, 363–366. doi:10.1038/318363a0

20. Archer M, Godthelp H, Hand SJ (1993) Early Eocene marsupial from Australia. In *Kaupia: Darmstadter beitrage zur naturgeschichte monument grube messel: Perspectives and Relationships. Part 2*. (Eds F Schrenk and K Ernst) pp. 193–200. Hessisches Landesmuseum Darmstadt, Darmstadt.

21. Archer M, Hand SJ, Godthelp H (1988) A new order of Tertiary zalambdodont marsupials. *Science* **239**, 1528–1531. doi:10.1126/science.239.4847.1528

22. Archer M, Hand SJ, Godthelp H (1994) *Australia's Lost World. Prehistoric Animals of Riversleigh*. Indiana University Press, Bloomington.

23. Archer M, Plane M, Pledge N (1978) Additional evidence for interpreting the Miocene *Obdurodon insignis* Woodburne and Tedford, 1975, to be a fossil platypus (Ornithorhynchidae: Monotremata) and a reconsideration of the status of *Ornithorhynchus agilis* De Vis, 1885. *Australian Zoologist* **20**, 9–19.

24. Archer M, Rich TH (1982) Results of the Ray E. Lemley expeditions. *Wakaleo alcootaensis* n. sp. (Thylacoleonidae, Marsupialia), a new marsupial lion from the Miocene of the Northern Territory with a consideration of early radiation in the family. In *Carnivorous Marsupials*. (Ed. M Archer) pp. 495–502. Royal Zoological Society of New South Wales, Sydney.

25. Arena DA (2004) The geological history and development of the terrain at the Riversleigh World Heritage Area during the Middle Tertiary. Unpublished PhD thesis. University of New South Wales, Sydney.

26. Arena DA, Black KH, Archer M, Hand SJ, Godthelp H, Creaser P (2014) Reconstructing a Miocene pitfall trap: recognition and interpretation of fossiliferous Cenozoic palaeokarst. *Sedimentary Geology* **304**, 28–43. doi:10.1016/j.sedgeo.2014.01.005

27. Attard MRG, Parr W, Wilson LAB, Archer M, Hand SJ, Rogers TL, Wroe S (2014) Virtual reconstruction and prey size preference in the mid Cenozoic thylacinid, *Nimbacinus dicksoni* (Thylacinidae, Marsupialia). *PLoS One* **9**(4), e93088. doi:10.1371/journal.pone.0093088

28. Australian Heritage Council (2012) *Australia's Fossil Heritage: A Catalogue of Important Australian Fossil Sites*. CSIRO Publishing, Melbourne.

29. Baird RF, Vickers-Rich P (1998) *Palaelodus* (Aves: Palaelodidae) from the Middle to Late Cainozoic of Australia. *Alcheringa* **22**, 135–151. doi:10.1080/03115519808619196

30. Balouet J-C (1991) The fossil vertebrate record of New Caledonia. In *Vertebrate Palaeontology of Australasia*. (Eds P Vickers-Rich, JM Monaghan, RF Baird and TH Rich) pp. 1383–1409. Monash University Publications Committee, Melbourne.

31. Balouet J-C, Buffetaut E (1987) *Mekosuchus inexpectatus*, n. g., n. sp., *Crocodilien nouveau de l'Holocène de Nouvelle Calédonie. Comptes Rendus de l'Académie des Sciences, Series II, Mecanique, Physique, Chimie, Sciences de l'Universitie, Paris* **304**, 853–856.

32. Balouet J-C, Olson SL (1989) Fossil birds from Late Quaternary deposits in New Caledonia. *Smithsonian Contributions to Zoology* **469**, 1–38. doi:10.5479/si.00810282.469

33. Barrie DJ (1990) Skull elements and associated remains of the Pleistocene boid snake *Wonambi naracoortensis*. *Memoirs of the Queensland Museum* **28**, 139–151.

34. Bartholomai A (1970) The extinct genus *Procoptodon* Owen (Marsupialia: Macropodidae) in Queensland. *Memoirs of the Queensland Museum* **15**, 213–233.

35. Bartholomai A, Molnar RE (1981) *Muttaburrasaurus*, a new iguanodont (Ornithischia: Ornithopoda) dinosaur from the Lower Cretaceous of Queensland. *Memoirs of the Queensland Museum* **20**, 319–349.

36. Bean LB (2006) The leptolepid fish *Cavenderichthys talbragarensis* (Woodward, 1895) from the Talbragar Fish Bed (Late Jurassic) near Gulgong, New South Wales. *Records of the Western Australian Museum* **23**, 43–76. doi:10.18195/issn.0312-3162.23(1).2006.043-076

37. Beck RMD (2009) Was the Oligo-Miocene Australian metatherian *Yalkaparidon* a 'mammalian woodpecker'? *Biological Journal of the Linnean Society. Linnean Society of London* **97**, 1–17. doi:10.1111/j.1095-8312.2009.01171.x

38. Beck RMD, Godthelp H, Weisbecker V, Archer M, Hand SJ (2008) Australia's oldest marsupial fossils and their biogeographical implications. *PLoS One* **3**(3), e1858. doi:10.1371/journal.pone.0001858

39. Beck RMD, Louys J, Brewer P, Archer M, Black KH, Tedford RH (2020) A new family of diprotodontian marsupials from the latest Oligocene of Australia and the evolution of wombats, koalas, and their relatives (Vombatiformes). *Scientific Reports* **10**, 9741. doi:10.1038/s41598-020-66425-8

40. Bell PR, Fanti F, Hart LJ, Milan LA, Craven SJ, Brougham T, Smith E (2019) Revised geology, age, and vertebrate diversity of the dinosaur-bearing Griman Creek Formation (Cenomanian), Lightning Ridge, New South Wales, Australia. *Palaeogeography, Palaeoclimatology, Palaeoecology* **514**, 655–671. doi:10.1016/j.palaeo.2018.11.020

41. Benson RBJ, Evans M, Smith AS, Sassoon J, Moore-Faye S, Ketchum HF, Forrest R (2013) A giant pliosaurid skull from the Late Jurassic of England. *PLoS One* **8**, e65989. doi:10.1371/journal.pone.0065989

42. Benson RBJ, Rich TH, Vickers-Rich P, Hall M (2012) Theropod fauna from southern Australia indicates high polar diversity and climate-driven dinosaur provinciality. *PLoS One* **7**, e37122. doi:10.1371/journal.pone.0037122

43. Berrell RW, Boisvert C, Trinajstic K, Siversson M, Alvarado-Ortega J, Cavin L, Salisbury SW, Kemp A (2020) A review of Australia's Mesozoic fishes. *Alcheringa* doi:10.1080/03115518.2019.1701078.

44. Berta A (2017) *The Rise of Marine Mammals: 50 Million Years of Evolution*. Johns Hopkins University Press, Baltimore.

45. Black K (1997) A new species of Palorchestidae (Marsupialia) from the Late Middle to early Late Miocene Encore Local Fauna, Riversleigh, northwestern Queensland. *Memoirs of the Queensland Museum* **41**, 181–186.

46. Black K (2010) *Ngapakaldia bonythoni* (Marsupialia, Diprotodontidae): new material from Riversleigh, northwestern Queensland, and a reassessment of the genus *Bematherium*. *Alcheringa* **34**, 471–492. doi:10.1080/03115511003793496

47. Black K, Archer M (1997) *Silvabestius*, a new genus and two new species of primitive zygomaturines (Marsupialia, Diprotodontidae) from Riversleigh, northwestern Queensland. *Memoirs of the Queensland Museum* **41**, 181–208.

48. Black K, Archer M (1997) *Nimiokoala* gen. nov. (Marsupialia, Phascolarctidae) from Riversleigh, northwestern Queensland, with a revision of *Litokoala*. *Memoirs of the Queensland Museum* **41**, 209–228.

49. Black KH, Archer M, Hand SJ (2012) New Tertiary koala (Marsupialia, Phascolarctidae) from Riversleigh, Australia, with a revision of phascolarctid phylogenetics, paleoecology, and paleobiodiversity. *Journal of Vertebrate Paleontology* **32**, 125–138. doi:10.1080/02724634.2012.626825

50. Black KH, Archer M, Hand SJ, Godthelp HJ (2010) First comprehensive analysis of cranial ontogeny in a fossil marsupial from a 15-million-year-old cave deposit in northern Australia. *Journal of Vertebrate Paleontology* **30**, 993–1011. doi:10.1080 /02724634.2010.483567

51. Black KH, Archer M, Hand SJ, Godthelp H (2012) The rise of Australian marsupials: a synopsis of biostratigraphic, phylogenetic, palaeoecologic and palaeobiogeographic understanding. In *Earth and Life: Global Biodiversity, Extinction Intervals and Biogeographic Perturbations through Time.* (Ed. JA Talent) pp. 983–1078. International Year of Planet Earth Series, Springer Verlag, Dordrecht.

52. Black KH, Archer M, Hand SJ, Godthelp H (2013) Revision of the diprotodontid marsupial genus *Neohelos*: systematics and biostratigraphy. *Acta Palaeontologica Polonica* **58**, 679–706.

53. Black KH, Camens AB, Archer M, Hand SJ (2012) Herds overhead: *Nimbadon lavarackorum* (Diprotodontidae), heavyweight marsupial herbivores in the Miocene forests of Australia. *PLoS One* **7**(11), e48213. doi:10.1371/journal. pone.0048213

54. Black KH, Hand SJ (2010) First crania and assessment of species boundaries in *Nimbadon* (Marsupialia: Diprotodontidae) from the Middle Miocene of Australia. *American Museum Novitates* **3678**, 1–60. doi:10.1206/666.1

55. Black KH, Louys J, Price GJ (2014) Understanding morphological variation in the extant koala as a framework for identification of species boundaries in extinct koalas (Phascolarctidae; Marsupialia). *Journal of Systematic Palaeontology* **12**, 237–264. doi:10.1080/14772019.2013.768304

56. Boast AP, Chapman B, Herrera MB, Worthy TH, Scofield RP, Tennyson AJD, Houde P, Bunce M, Cooper A, Mitchell KJ (2019) Mitochondrial genomes from New Zealand's extinct adzebills (Aves: Aptornithidae: Aptornis) support a sister-taxon relationship with the Afro-Madagascan Sarothruridae. *Diversity* **11**, 24. doi:10.3390/ d11020024

57. Boisvert CA, Mark-Kurik E, Ahlberg PE (2008) The pectoral fin of *Panderichthys* and the origin of digits. *Nature* **456**, 636–638. doi:10.1038/ nature07339

58. Boles WE (1995) The world's oldest songbird. *Nature* **374**, 21–22. doi:10.1038/374021b0

59. Boles WE (1995) A preliminary analysis of the Passeriformes from Riversleigh, northwestern Queensland, Australia, with the description of a new species of lyrebird. *Courier Forschungsinstitut Senckenberg* **181**, 163–170.

60. Boles WE (1997) Fossil songbirds (Passeriformes) from the Early Eocene of Australia. *Emu* **97**, 43–50. doi:10.1071/MU97004

61. Boles WE (1999) Early Eocene shorebirds (Aves: Charadriiformes) from the Tingamarra Local Fauna, Murgon, Queensland, Australia. *Records of the Western Australian Museum* **57**(supp.), 229–238.

62. Boles WE (2005) A review of the Australian fossil storks of the genus *Ciconia* (Aves: Ciconiidae), with the description of a new species. *Records of the Australian Museum* **57**, 165–178. doi:10.3853 /j.0067-1975.57.2005.1440

63. Boles WE (2008) Systematics of the fossil Australian giant megapodes *Progura* (Aves: Megapodiidae). *Oryctos* **7**, 195–295.

64. Brammall J, Archer M (1999) Living and extinct petaurids, acrobatids, tarsipedids and burramyids (Marsupialia): relationships and diversity through time. *Australian Mammalogy* **21**, 24–25.

65. Brathwaite DH (1992) Notes on the weight, flying ability, habitat, and prey of Haast's eagle (*Harpagornis moorei*). *Notornis* **39**, 239–247.

66. Brochu CA (2001) Crocodylian snouts in space and time: phylogenetic approaches toward adaptive radiation. *American Zoologist* **41**, 564–585. doi:10.1093/icb/41.3.564

67. Bunce M, Szulkin M, Lerner HRL, Barnes I, Shapiro B, Cooper A, Holdaway RN (2005) Ancient DNA provides new insights into the evolutionary history of New Zealand's extinct giant eagle. *PLoS Biology* **3**, e9. doi:10.1371/journal.pbio.0030009

68. Burrow CJ, Young GC (1999) An articulated teleostome fish from the Late Silurian (Ludlow) of Victoria, Australia. *Records of the Western Australian Museum* **57**(supp.), 1–14.

69. Butler K, Travouillon KJ, Price GJ, Archer M, Hand SJ (2018) Revision of Oligo-Miocene kangaroos, *Ganawamaya* and *Nambaroo* (Marsupialia: Macropodiformes). *Palaeontologia Electronica* **21.1.8A**, 1–58. doi:10.26879/747

70. Caldwell MW, Holmes R, Bell GL Jr, Wiffen J (2005) An unusual tylosaurine mosasaur from New Zealand: a new skull of *Taniwhasaurus oweni* (Lower Haumurian: Upper Cretaceous). *Journal of Vertebrate Paleontology* **25**, 393–401. doi:10.1671/0272-4634(2005)025[0393:AUTMFN]2.0.CO;2

71. Campbell CR (1973) A new species of *Troposodon bartholomaii*; from the Early Pleistocene Kanunka Fauna, South Australia (Macropodinae; Marsupialia). *Records of the South Australian Museum* **16**, 1–18.

72. Campbell KSW, Phuoc L (1983) A Late Permian actinopterygian fish from Australia. *Palaeontology* **26**, 33–70.

73. Choo B (2011) Revision of the actinopterygian genus *Mimipiscis* (=*Mimia*) from the Upper Devonian Gogo Formation of Western Australia and the interrelationships of the early Actinopterygii. *Earth and Environmental Science Transactions of the Royal Society of Edinburgh* **11**, 1–28.

74. Choo B (2015) A new species of the Devonian actinopterygians *Moythomasia* from Bergisch Gladbach, Germany, and fresh observations on *M. durgaringa* from the Gogo Formation of Western Australia. *Journal of Vertebrate Paleontology* **35**, e952817. doi:10.1080/02724634.2015.952817

75. Clack JA (2012) *Gaining Ground: The Origin and Evolution of Tetrapods*, 2nd edn. Indiana University Press, Bloomington.

76. Clemens WA, Plane M (1974) Mid-Tertiary Thylacoleonidae (Marsupialia, Mammalia). *Journal of Paleontology* **48**, 652–660.

77. Clites EC, Droser ML, Gehling JG (2012) The advent of hard-part structural support among Ediacra biota: Ediacaran harbinger of a Cambrian mode of body construction. *Geology* **40**, 307–310. doi:10.1130/G32828.1

78. Coates MI, Clack JA (1990) Polydactyly in the earliest known tetrapod limbs. *Nature* **347**, 66–69. doi:10.1038/347066a0

79. Colbert EH, Merrilees DM (1967) Cretaceous dinosaur footprints from Western Australia. *Journal of the Royal Society of Western Australia* **50**, 21–25.

80. Conway Morris S, Caron J (2014) A fish from the Cambrian of North America. *Nature* **512**, 419–422. doi:10.1038/nature13414

81. Cooke BN (1992) Primitive macropodids from Riversleigh, north-western Queensland. *Alcheringa* **16**, 201–217. doi:10.1080/03115519208619119

82. Cooke BN (1997) Biostratigraphic implications of fossil kangaroos at Riversleigh, northwestern Queensland. *Memoirs of the Queensland Museum* **41**, 295–302.

83. Cooke BN (2000) Cranial remains of a new species of balbarine kangaroo (Marsupalia: Macropodoidea) from the Oligo-Miocene freshwater limestone deposits of Riversleigh World Heritage Area, northern Australia. *Journal of Paleontology* **74**, 317–326. doi:10.1666/0022-3360(2000)074<0317:CROANS>2.0.CO;2

84. Cooke BN, Travouillon KJ, Archer M, Hand SJ (2015) *Ganguroo robustiter* sp. nov. (Macropodidae, Marsupialia): a middle to early Late Miocene basal macropodid from Riversleigh World Heritage Area, Australia. *Journal of Vertebrate Paleontology* **35**, e956879. doi:10.1080/02724634.2015.956879

85. Cookson IC (1935) On plant remains from the Silurian of Victoria, Australia that extend and connect floras hitherto described. *Philosophical Transactions of the Royal Society of London. Series B, Biological Sciences* **225**, 127–148.

86. Cosgriff JW (1965) A new genus of Temnospondyli from the Triassic of Western Australia. *Journal of the Royal Society of Western Australia* **48**, 65–90.

87. Cosgriff JW, Garbutt NK (1972) *Erythorobatrachus noonkanbahensis*, a trematosaurid species from the Blina Shale. *Journal of the Royal Society of Western Australia* **55**, 5–18.

88. Crosby K (2007) Rediagnosis of the fossil species assigned to *Strigocuscus* (Marsupialia, Phalangeridae), with description of a new genus and three new species. *Alcheringa* **31**, 33–58. doi:10.1080/03115510601123619

89. Cruickshank ARI, Fordyce RE (2002) A new marine reptile (Sauropterygia) from New Zealand: further evidence for a Late Cretaceous austral radiation of cryptoclidid plesiosaurs. *Palaeontology* **45**, 557–575. doi:10.1111/1475-4983.00249

90. Davies NS, Sansom IJ, Nicoll RS, Ritchie A (2011) Ichnofacies of the Stairway Sandstone fish-fossil beds (Middle Ordovician, Northern Territory, Australia). *Alcheringa* **35**, 553–569. doi:10.1080/03115518.2011.557565

91. Davis AC, Archer M (1997) *Palorchestes azael* (Mammalia, Palorchestidae) from the Late Pleistocene Terrace Site Local Fauna, Riversleigh, northwestern Queensland. *Memoirs of the Queensland Museum* **41**, 315–320.

92. Dawson L (1985) Marsupial fossils from Wellington Caves, New South Wales: the historic and scientific significance of the collections in the Australian Museum, Sydney. *Records of the Australian Museum* **37**, 55–69. doi:10.3853/j.0067-1975.37.1985.335

93. Dawson L, Augee ML (1997) The Late Quaternary sediments and fossil vertebrate fauna from Cathedral Cave, Wellington Caves, New South Wales. *Proceedings of the Linnean Society of New South Wales* **117**, 51–78.

94. De Deckker P (1982) Non-marine ostracods from two Quaternary profiles at Pulbeena and Mowbray Swamps, Tasmania. *Alcheringa* **6**, 249–274. doi:10.1080/03115518208619636

95. de Muizon C (1994) Are the squalodonts related to the platanistoids? *Proceedings of the San Diego Society of Natural History* **29**, 135–146.

96. Delcourt R, Grillo ON (2018) Tyrannosauroids from the Southern Hemisphere: implications for biogeography, evolution, and taxonomy. *Palaeogeography, Palaeoclimatology, Palaeoecology* **511**, 379–387. doi:10.1016/j.palaeo.2018.09.003

97. Delsuc F, Brinkmann H, Chourrout D, Philippe H (2006) Tunicates and not cephalochordates are the closest living relative of vertebrates. *Nature* **439**, 965–968. doi:10.1038/nature04336

98. Dennis KD, Miles RS (1979) Eubrachythoracid arthrodires with tubular rostral plates from Gogo, Western Australia. *Zoological Journal of the Linnean Society* **67**, 297–328. doi:10.1111/j.1096-3642.1979.tb01118.x

99. Dias-da-Silva S, Marsicano C (2011) Phylogenetic reappraisal of Rhytidosteidae (Stereospondyli: Trematosauria), temnospondyl amphibians from the Permian and Triassic. *Journal of Systematic Palaeontology* **9**, 305–325. doi:10.1080/14772019.2010.492664

100. Djokic T, van Kranendonk MJ, Campbell A, Walter MR, Ward CR (2017) Earliest signs of life on land preserved in ca. 3.5 Ga hot spring deposits. *Nature Communications* **8**, doi:10.1038/ncomms15263.

101. Dobson JS, Zdenek CN, Hay C, Violette A, Fourmy R, Cochran C, Fry BG (2019) Varanid lizard venoms disrupt the clotting ability of human fibrinogen through destructive cleavage. *Toxins* **11**(5), 255–267. doi:10.3390/toxins11050255

102. Douglas J, Holmes F (2006) The *Baragwanathia* story: an update. *The Fossil Collector* **77**, 9–26.

103. Droser ML, Gehling JG (2008) Synchronous aggregate growth in an abundant new Ediacran Tubular organism. *Science* **319**, 1660–1662. doi:10.1126/science.1152595

104. Dziewa TJ (1980) Early Triassic osteichthyans from the Triassic Knocklofty Formation of Tasmania. *Papers and Proceedings of the Royal Society of Tasmania* **114**, 145–160. doi:10.26749/rstpp.114.145

105. Edgecombe GD, García-Bellido DC, Paterson JR (2011) A new leanchoiliid megacheiran arthropod from the lower Cambrian Emu Bay Shale, South Australia. *Acta Palaeontologica Polonica* **56**, 385–400. doi:10.4202/app.2010.0080

106. Ellis R (2003) *Sea Dragons: Predators of the Prehistoric Oceans*. University Press of Kansas, Lawrence.

107. Ezcurra MD (2014) The osteology of the basal archosauromorph *Tasmaniosaurus triassicus* from the Lower Triassic of Tasmania, Australia. *PLoS One* **9**(1), e86864. doi:10.1371/journal. pone.0086864

108. Ezcurra M (2016) The phylogenetic relationships of basal archosauromorphs, with an emphasis on the systematics of proterosuchian archosauriforms. *PeerJ* **4**, e1778. doi:10.7717/ peerj.1778

109. Faith JT, O'Connell JF (2011) Revisiting the late Pleistocene mammal extinction record at Tight Entrance Cave, southwestern Australia. *Quaternary Research* **76**, 397–400. doi:10.1016/j. yqres.2011.08.001

110. Fischer V, Benson RBJ, Druckenmiller PS, Ketchum HF, Bardet N (2018) The evolutionary history of polycotylid plesiosaurians. *Royal Society Open Science* **5**, 172177. doi:10.1098/ rsos.172177

111. Fitzgerald EMG (2006) A bizarre new toothed mysticete (Cetacea) from Australia and the early evolution of baleen whales. *Proceedings. Biological Sciences* **273**, 2955–2963. doi:10.1098/ rspb.2006.3664

112. Fitzgerald EMG (2010) The morphology and systematics of *Mammalodon colliveri* (Cetacea: Mysticeti), a toothed mysticete from the Oligocene of Australia. *Zoological Journal of the Linnean Society* **158**, 367–476. doi:10.1111/j.1096-3642.2009.00572.x

113. Fitzgerald EMG (2016) A late Oligocene waipatiid dolphin (Odontoceti: Waipatiidae) from Victoria, Australia. *Memoirs of Museum Victoria* **74**, 117–136.

114. Fitzgerald EMG, Park T, Worthy TH (2012) First giant bony-toothed bird (Pelagornithidae) from Australia. *Journal of Vertebrate Paleontology* **32**, 971–974. doi:10.1080/02724634.2012.664596

115. Flannery TF (1994) The fossil land mammal record of New Guinea: a review. *Science in New Guinea* **20**, 39–48.

116. Flannery TF (1999) The Pleistocene fauna of Kelangurr Cave, central montane Irian Jaya, Indonesia. *Records of the Western Australian Museum* (supp. 57), 341–350.

117. Flannery TF, Archer M (1985) *Palorchestes*: large and small palorchestids. In *Kadimakara: Extinct Vertebrates of Australia*. (Eds PV Rich, GF van Tets and F Knight) pp. 234–239. Pioneer Design Studio, Melbourne.

118. Flannery T, Archer M (1987) *Strigocuscus reidi* and *Trichosurus dicksoni*, two new phalangerids (Marsupialia: Phalangeridae) from the Miocene of northwestern Queensland. In *Possums and Opossums: Studies in Evolution*. (Ed. M Archer) pp. 527–536. Surrey Beatty & Sons and Royal Zoological Society of New South Wales, Sydney.

119. Flannery TF, Archer M, Rich TH, Jones R (1995) A new family of monotremes from the Cretaceous of Australia. *Nature* **377**, 418–420. doi:10.1038/377418a0

120. Flannery TF, Hoch E, Aplin K (1989) Macropodines from the Pliocene Otibanda Formation, Papua New Guinea. *Alcheringa* **13**, 145–152. doi:10.1080/03115518908619048

121. Flannery TF, Mountain M-J, Aplin K (1983) Quaternary kangaroos (Macropodidae: Marsupialia) from Nombe Rock Shelter, Papua New Guinea, with comments on the nature of megafaunal extinction in the New Guinea highlands. *Proceedings of the Linnean Society of New South Wales* **107**, 75–97.

122. Flannery TF, Plane M (1986) A new late Pleistocene diprotodontid (Marsupialia) from Pureni, Southern Highlands Province, Papua New Guinea. *BMR Journal of Australian Geology and Geophysics* **10**, 65–76.

123. Fletcher TL, Greenwood DR, Moss PT, Salisbury SW (2014) Paleoclimate of the late Cretaceous (Cenmomanian-Turonian) portion of the Winton Formation, central-western Queensland, Australia: new observations based on clamp and biometric analysis. *Palaios* **29**, 121–128. doi:10.2110/palo.2013.080

124. Fordyce RE (1982) A review of Australian fossil Cetacea. *Memoirs of the National Museum Victoria* **43**, 43–58. doi:10.24199/j. mmv.1982.43.04

125. Fordyce RE (1983) Rhabdosteid dolphins (Mammalia: Cetacea) from the middle Miocene, Lake Frome area, South Australia. *Alcheringa* **7**, 27–40. doi:10.1080/03115518308619631

126. Fordyce RE (1984) Evolution and zoogeography of cetaceans in Australia. In *Vertebrate Zoogeography and Evolution in Australasia.* (Eds M Archer and G Clayton) pp. 929–948. Hesperian Press, Perth.

127. Fordyce RE (1994) *Waipatia maerewhenua*, new genus and new species (Waipatiidae, new family), an archaic Late Oligocene dolphin (Cetacea: Odontoceti: Platanistoidea) from New Zealand. In *Contributions in Marine Mammal Paleontology Honoring Frank C. Whitmore, Jr.* (Eds A Berta and T Deméré) pp. 147–176. *Proceedings of the San Diego Museum of Natural History* 29.

128. Fraser RA, Wells RT (2006) Palaeontological excavation and taphonomic investigation of the late Pleistocene fossil deposit in Grant Hall, Victoria Fossil Cave, Naracoorte, South Australia. *Alcheringa Special Issue* **1**, 147–161.

129. Fry B, Wroe S, Teeuwisse W, van Osch MJP, Moreno K, Ingle J, McHenry C, Ferrara T, Clausen P, Scheib H, Winter KL, Greisman L, Roelants K, van der Weerd L, Clemente CJ, Giannakis E, Hodgson WC, Luz S, Martelli P, Krishnasamy K, Kochva E, Kwok HF, Scanlon D, Karas J, Citron DM, Goldstein EJC, McNaughtan JE, Norman JA, et al. (2009) A central role for venom in predation by *Varanus komodoensis* (Komodo Dragon) and the extinct giant *Varanus* (*Megalania*) *priscus. Proceedings of the National Academy of Sciences of the United States of America* **106**(22), 8969–8974. doi:10.1073/pnas.0810883106

130. Gaffney ES (1983) Cranial morphology of the extinct horned turtle, *Meiolania platyceps*, from the Pleistocene of Lord Howe Island. *Bulletin of the American Museum of Natural History* **175**, article 4.

131. Gaffney ES (1985) The cervical and caudal vertebrae of the cryptodiran turtle, *Meiolania platyceps*, from the Pleistocene of Lord Howe Island, Australia. *American Museum Novitates* **2805**, 1–29.

132. Gaffney ES (1992) *Ninjemys*, a new name for "*Meiolania*" *oweni* (Woodward), a horned turtle from the Pleistocene of Queensland. *American Museum Novitates* **3049**, 1–10.

133. Gaffney ES (1996) The postcranial morphology of *Meiolania platyceps* and a review of the Meiolaniidae. *Bulletin of the American Museum of Natural History* **229**, 1–166.

134. Gaffney ES, Archer M, White A (1989) Chelid turtles from the Miocene freshwater limestones of Riversleigh Station, northwestern Queensland, Australia. *American Museum Novitates* **2959**, 1–10.

135. Gaffney ES, Kool L, Brinkman DB, Rich TH, Vickers-Rich P (1998) *Otwayemys*, a new cryptodiran turtle from the Early Cretaceous of Australia. *American Museum Novitates* **3233**, 1–28.

136. Gaffney ES, McNamara G (1990) A meiolaniid turtle from the Pleistocene of northern Queensland. *Memoirs of the Queensland Museum* **28**, 107–113.

137. García-Bellido DC, Paterson JR, Edgecombe GD, Jago JB, Gehling JG, Lee MSY (2009) The bivalved arthropods *Isoxys* and *Tuzoia* with soft-part preservation from the lower Cambrian Emu Bay Shale Lagerstätte (Kangaroo Island, Australia). *Palaeontology* **52**, 1221–1241. doi:10.1111/j.1475-4983.2009.00914.x

138. Garratt MJ, Rickards BR (1984) Graptolite biostratigraphy of early land plants from Victoria, Australia. *Proceedings of the Yorkshire Geological Society* **44**, 377–384. doi:10.1144/pygs.44.4.377

139. Garvey JM, Johanson Z, Warren A (2005) Redescription of the pectoral fin and vertebral column of the rhizodontid fish *Barameda decipiens* from the Lower Carboniferous of Australia. *Journal of Vertebrate Paleontology* **25**, 8–18. doi:10.1671/0272-4634(2005)025[0008:ROTPFA]2.0.CO;2

140. Gillespie A (1997) *Priscileo roskellyae* sp. nov. (Thylacoleonidae, Marsupialia) from the Oligocene-Miocene of Riversleigh, northwestern Queensland. *Memoirs of the Queensland Museum* **41**, 321–327.

141. Gillespie AK, Archer M, Hand SJ (2016) A tiny new marsupial lion (Marsupialia, Thylacoleonidae) from the Early Miocene of Australia. *Palaeontologia Electronica* **19**(229A), 1–25. doi:10.26879/632

142. Gillespie AK, Archer M, Hand SJ (2017) A new Oligo-Miocene marsupial lion from Australia and revision of the family Thylacoleonidae. *Journal of Systematic Palaeontology* **17**, 59–89. doi:10.1080/14772019.2017.1391885

143. Gillespie AK, Archer M, Hand SJ (2019) *Lekaneleo*, a new genus of marsupial lion (Marsupialia, Thylacoleonidae) from the Oligocene–Miocene of Australia, and the craniodental morphology of *L. roskellyae*, comb. nov. *Journal of Vertebrate Paleontology* **39**, e1703722. doi:10.1080/02724634.2019.1703722

144. Gillespie AK, Archer M, Hand SJ, Black K (2014) New material referable to *Wakaleo* (Marsupialia: Thylacoleonidae) from the Riversleigh World Heritage Area, northwestern Queensland: revising species boundaries and distributions in Oligo-Miocene marsupial lions. *Alcheringa* **38**, 513–527. doi:10.1080/03115518.2014.908268

145. Gillespie R, Camens AB, Worthy TH, Rawlence NJ, Reid C, Bertuch F, Levchenko V, Cooper A (2012) Man and megafauna in Tasmania: closing the gap. *Quaternary Science Reviews* **37**, 38–47. doi:10.1016/j.quascirev.2012.01.013

146. Gillespie R, Wood R, Fallon S, Stafford TWJ, Southon J (2015) New 14C dates for Spring Creek and Mowbray Swamp megafauna: XAD-2 processing. *Archaeology in Oceania* **50**, 43–48. doi:10.1002/arco.5045

147. Glauert L (1910) The Mammoth Cave. *Records of the Western Australian Museum and Art Gallery* **1**, 11–36.

148. Godthelp H, Archer M, Cifelli R, Hand SJ, Gilkeson CF (1992) Earliest known Australian Tertiary mammal fauna. *Nature* **356**, 514–516. doi:10.1038/356514a0

149. Godthelp H, Wroe S, Archer M (1999) A new marsupial from the early Eocene Tingamarra Local Fauna of Murgon, southeastern Queensland: a prototypical Australian marsupial? *Journal of Mammalian Evolution* **6**, 289–313. doi:10.1023/A:1020517808869

150. Greenwood RM, Atkinson IAE (1977) Evolution of divaricating plants in New Zealand in relation to moa browsing. *New Zealand Ecological Society Proceedings* **24**, 21–23.

151. Gurovich Y, Travouillon KJ, Beck RMD, Muirhead J, Archer M (2013) Biogeographical implications of a new mouse-sized fossil bandicoot (Marsupialia: Peramelemorphia) occupying a dasyurid-like ecological niche across Australia. *Journal of Systematic Palaeontology* **12**, 265–290. doi:10.1080/14772019.2013.776646

152. Haast J (1872) Notes on *Harpagornis moorei*, an extinct gigantic bird of prey, containing description of femur, ungual phalanges, and rib. *Transactions and Proceedings of the New Zealand Institute* **4**, 192–196.

153. Hammer WR, Hickerson WJ (1994) A crested theropod dinosaur from Antarctica. *Science* **264**, 828–830. doi:10.1126/science.264.5160.828

154. Hammer WR, Smith MD (2008) A tritylodont postcanine from the Hanson Formation of Antarctica. *Journal of Vertebrate Paleontology* **28**, 269–273. doi:10.1671/0272-4634(2008)28[269:ATPFTH]2.0.CO;2

155. Hand SJ (1996) New Miocene and Pliocene megadermatids (Mammalia; Microchiroptera) from Australia, with comments on broader aspects of megadermatid evolution. *Geobios* **29**, 365–377. doi:10.1016/S0016-6995(96)80038-6

156. Hand SJ (1997) New Miocene leaf-nosed bats (Microchiroptera: Hipposideridae) from Riversleigh Station, Queensland. *Memoirs of the Queensland Museum* **41**, 335–349.

157. Hand SJ (1998) *Riversleigha williamsi* n. gen. et n. sp., a large Miocene hipposiderid from Riversleigh, Queensland. *Alcheringa* **22**, 259–276. doi:10.1080/03115519808619204

158. Hand SJ (1998) *Xenorhinos*, a new genus of Old World leaf-nosed bats (Microchiroptera: Hipposideridae) from the Australian Miocene. *Journal of Vertebrate Paleontology* **18**, 430–439. doi:10.1080/02724634.1998.10011070

159. Hand S, Archer M (2005) A new hipposiderid genus (Microchiroptera) from an early Miocene bat community in Australia. *Palaeontology* **48**, 371–383. doi:10.1111/j.1475-4983.2005.00444.x

160. Hand S, Archer M, Godthelp H (2005) Australian Oligo-Miocene mystacinids (Microchiroptera): upper dentition, new taxa and divergence of New Zealand species. *Geobios* **38**, 339–352. doi:10.1016/j.geobios.2003.11.005

161. Hand S, Archer M, Rich T, Pledge N (1993) *Nimbadon*, a new genus and three species of Tertiary zygomaturines (Marsupialia, Diprotodontidae) from northern Australia, with a reassessment of *Neohelos. Memoirs of the Queensland Museum* **33**, 193–210.

162. Hand SJ, Beck RMD, Archer M, Simmons NB, Gunnell GF, Scofield RP, Tennyson AJD, De Pietri VL, Salisbury SW, Worthy TH (2018) A new, large-bodied omnivorous bat (Noctilionoidea: Mystacinidae) reveals lost morphological and ecological diversity since the Miocene in New Zealand. *Scientific Reports* **8**, 235. doi:10.1038/s41598-017-18403-w

163. Hand SJ, Novacek MJ, Godthelp H, Archer M (1994) First Eocene bat from Australia. *Journal of Vertebrate Paleontology* **14**, 375–381. doi:10.1080/02724634.1994.10011565

164. Hand SJ, Worthy TH, Archer M, Worthy JP, Tennyson AJD, Scofield RP (2013) Miocene mystacinids (Chiroptera: Noctilionoidea) indicate a long history for endemic bats in New Zealand. *Journal of Vertebrate Paleontology* **33**, 1442–1448. doi:10.1080/02724634.2013.775950

165. Hand SJ, Weisbecker V, Beck RMD, Archer M, Godthelp H, Tennyson AJD, Worthy TH (2009) Bats that walk: a new evolutionary hypothesis for the terrestrial behaviour of New Zealand's endemic mystacinids. *BMC Evolutionary Biology* **9**, 169. doi:10.1186/1471-2148-9-169

166. Hart LJ, Bell PR, Smith ET, Salisbury SW (2019) *Isisfordia molnari* sp. nov., a new basal eusuchian from the mid-Cretaceous of Lightning Ridge, Australia. *PeerJ* **7**, e7166. doi:10.7717/peerj.7166

167. Hatcher L (2009) *Palaeontology of Mammoth Cave.* http://www.megafauna.com.au/upload/pages/research/palaeo-mammoth-school-fact-sheet-updated-4-4-11.pdf.

168. Hecht M, Archer M (1977) The presence of xiphodont crocodiles in the Tertiary and Pleistocene of Australia. *Alcheringa* **1**, 383–385. doi:10.1080/03115517708527772

169. Hector J (1874) On the fossil Reptilia of New Zealand. *Transactions and Proceedings of the New Zealand Institute* **6**, 333–358.

170. Helgen KM, Wells RT, Kear BP, Gerdtz WR, Flannery TF (2006) Ecological and evolutionary significance of sizes of giant extinct kangaroos. *Australian Journal of Zoology* **54**(4), 293–303. doi:10.1071/ZO05077

171. Herne M (2009) Postcranial osteology of *Leaellynasaura amicagraphica* (Dinosauria; Ornithischia) from the Early Cretaceous of southeastern Australia. *Journal of Vertebrate Paleontology* **29**, 33A.

172. Herne MC, Tait AM, Salisbury SW (2016) Sedimentological reappraisal of the *Leaellynasaura amicagraphica* (Dinosauria, Ornithopoda) holotype locality in the lower Cretaceous of Victoria, Australia, with taphonomic implications for the taxon. *New Mexico Museum of Natural History and Science Bulletin* **71**, 121–148.

173. Hiller N, Fordyce RE (compilers) (2009) New Zealand fossil reptiles. In Phylum Chordata: lancelets, fishes, amphibians, reptiles, birds, mammals. (Auth. CM King, CD Roberts, BD Bell, RE Fordyce, RS Nicoll, TH Worthy, CD Paulin, RA Hitchmough, IW Keyes, AN Baker, AL Stewart, N Hiller, RM McDowall, RN Holdaway, RP McPhee, WW Schwarzhans, AJD Tennyson, S Rust and I McCadie) p. 543. In *2009: New Zealand Inventory of Biodiversity. Volume 1. Kingdom Animalia. Radiata, Lophotrochozoa, Deuterostomia.* (Ed. DP Gordon) pp. 431–551. Canterbury University Press, Christchurch.

174. Hiller N, Mannering AA, Jones CM, Cruickshank ARI (2005) The nature of *Mauisaurus haasti* Hector, 1874 (Reptilia: Plesiosauria). *Journal of Vertebrate Paleontology* **25**, 588–601. doi:10.1671/0272-4634(2005)025[0588:TNOMHH]2.0.CO;2

175. Hocknull SA, Piper PJ, van den Bergh GD, Due RA, Morwood MJ, Kurniawan I (2009) Dragon's paradise lost: palaeobiogeography, evolution and extinction of the largest-ever terrestrial lizards (Varanidae). *PLoS One* **4**, e7241. doi:10.1371/journal.pone.0007241

176. Hocknull SA, White MA, Tischler TR, Cook AG, Calleja ND, Sloan T, Elliott DA (2009) New mid-Cretaceous (Latest Albian) dinosaurs from Winton, Queensland, Australia. *PLoS One* **4**(7), e6190. doi:10.1371/journal.pone.0006190

177. Hocknull SA, Wilkinson M, Lawrence RA, Konstantinov V, Mackenzie S, Mackenzie R (2021) A new giant sauropod, *Australotitan cooperensis* gen. et sp. nov., from the mid-Cretaceous of Australia. *PeerJ* **9**, e11317. doi:10.7717/peerj.11317

178. Holdaway RN (1999) Introduced predators and avifaunal extinction in New Zealand. In *Extinctions in Near Time: Causes, Contexts, and Consequences.* (Ed. RDE MacPhee) pp. 189–238. Kluwer Academic and Plenum Press, New York.

179. Holdaway RN, Worthy TH (1997) A reappraisal of the Late Quaternary fossil vertebrates of Pyramid Valley Swamp, North Canterbury, New Zealand. *New Zealand Journal of Zoology* **24**, 69–121. doi:10.1080/03014223.1997.9518107

180. Holland T, Warren A, Johanson Z, Long J, Parker K, Campbell JM (2007) A new species of *Barameda* (Rhizodontida) and heterochrony in the rhizodontid pectoral fin. *Journal of Vertebrate Paleontology* **27**, 295–315. doi:10.1671/0272-4634(2007)27[295:ANSOBR]2.0.CO;2

181. Holtz TR Jr (1994) The phylogenetic position of the Tyrannosauridae: implications for theropod systematics. *Journal of Paleontology* **68**, 1100–1117. doi:10.1017/S0022336000026706

182. Hou X-G, Aldridge RJ, Siveter DJ, Siveter DJ, Feng X-H (2002) New evidence on the anatomy and phylogeny of the earliest vertebrates. *Proceedings. Biological Sciences* **269**, 1865–1869. doi:10.1098/rspb.2002.2104

183. Howie AA (1972) A brachyopid labyrinthodont from the Lower Trias of Queensland. *Proceedings of the Linnean Society of New South Wales* **9**, 268–277.

184. Hutchinson MN (1992) Origins of the Australian scincid lizards: a preliminary report on the skinks of Riversleigh. *The Beagle: Records of the Northern Territory Museum of Arts and Sciences* **9**, 61–69. doi:10.5962/p.263118

185. Irwin G, Worthy TH, Best S, Hawkins S, Carpenter J, Matararaba S (2011) Further investigations at the Naigani Lapita site (VL 21/5), Fiji: excavation, radiocarbon dating and palaeofaunal extinction. *Journal of Pacific Archaeology* **2**, 66–78.

186. *Jan Juc Fossils.* https://museumsvictoria.com.au/media-releases/victorian-fossil-find-uncovers-prehistoric-leftovers-of-colossal-shark-feast/

187. Janis CM, Buttrill K, Figuerido B (2014) Locomotion in extinct giant kangaroos: were sthenurines hop-less monsters? *PLoS One* **9**(10), e109888. doi:10.1371/journal.pone.0109888

188. Janis CM, Damuth J, Travouillon KJ, Figueirido B, Hand SJ, Archer M (2016) Palaeoecology of Oligo-Miocene macropodoids determined from craniodental and calcaneal data. *Memoirs of the Museum of Victoria* **73**, 200–232.

189. Johanson Z (1997) New *Remigolepis* (Placodermi, Antiarchi) from Canowindra, New South Wales, Australia. *Geological Magazine* **134**, 813–846. doi:10.1017/S0016756897007838

190. Johanson Z (1998) The Upper Devonian *Bothriolepis* (Antiarchi, Placodermi) from near Canowindra, New South Wales, Australia. *Records of the Australian Museum* **50**, 315–348. doi:10.3853/j.0067-1975.50.1998.1289

191. Johanson Z, Ahlberg PE (1997) A new tristichopterid (Osteolepiformes: Sarcopterygii) from the Mandagery Sandstone (Late Devonian, Famennian) near Canowindra, NSW, Australia. *Transactions of the Royal Society of Edinburgh. Earth Sciences* **88**, 39–68. doi:10.1017/S0263593300002303

192. Johanson Z, Ahlberg PE (1998) A complete primitive rhizodonts from Australia. *Nature* **394**, 569–573. doi:10.1038/29058

193. Johanson Z, Ahlberg PE (2001) Devonian rhizodontids (Sarcopterygii; Tetrapodomorpha) from East Gondwana. *Transactions of the Royal Society of Edinburgh. Earth Sciences* **92**, 43–74. doi:10.1017/S0263593300000043

194. Johanson Z, Turner S, Warren A (2000) First East Gondwanan record of *Strepsodus* (Sarcopterygii, Rhizodontida) from the Lower Carboniferous Ducabrook Formation, central Queensland, Australia. *Geodiversitas* **22**, 161–169.

195. Jones MEH, Tennyson AJD, Worthy JP, Evans SE, Worthy TH (2009) A sphenodontine (Rhynchocephalia) from the Miocene of New Zealand and palaeobiogeography of the tuatara (*Sphenodon*). *Proceedings. Biological Sciences* **276**, 1385–1390. doi:10.1098/rspb.2008.1785

196. Joseph-Ouni M, McCord WP, Cann J, Smales I, Freeman A, Sadlier R, Couper P, White A, Amey A (2020) The relics of Riversleigh: re-examination of the fossil record of *Elseya* (Testudines: Chelidae) with description of a new extant species from the Gulf of Carpentaria Drainages, Queensland, Australia. *The Batagar Monographs* **3**, 7–69.

197. Kear BP (2005) A new elasmosaurid plesiosaur from the Lower Cretaceous of Queensland, Australia. *Journal of Vertebrate Paleontology* **25**, 792–805. doi:10.1671/0272-4634(2005)025[0792:ANEPFT]2.0.CO;2

198. Kear BP (2006) Marine reptiles from the Lower Cretaceous of South Australia: elements of a high latitude cold water assemblage. *Palaeontology* **49**, 837–856. doi:10.1111/j.1475-4983.2006.00569.x

199. Kear BP (2007) Taxonomic clarification of the Australian elasmosaurid genus *Eromangasaurus*, with reference to other austral *elasmosaur* taxa. *Memoirs of the Queensland Museum* **27**, 241–246.

200. Kear BP, Cooke BN, Archer M, Flannery TF (2007) Implications of a new species of the Oligo-Miocene kangaroo (Marsupialia: Macropodoidea) *Nambaroo*, from the Riversleigh World Heritage Area, Queensland. *Journal of Paleontology* **81**, 1147–1167. doi:10.1666/04-218.1

201. Kear BP, Hamilton-Bruce RJ (2011) *Dinosaurs in Australia: Mesozoic Life from the Southern Continent*. CSIRO Publishing, Melbourne.

202. Kear BP, Schroeder NI, Lee MSY (2006) An archaic crested plesiosaur in opal from the Lower Cretaceous high-latitude deposits of Australia. *Biology Letters* **2**, 615–619. doi:10.1098/rsbl.2006.0504

203. Kemp A (1997) A revision of Australian Mesozoic and Cenozoic lungfish of the family Neoceratodontidae (Osteichthyes: Dipnoi), with a description of four new species. *Journal of Paleontology* **71**, 713–733. doi:10.1017/S0022336000040166

204. Kemp A, Berrell RW (2020) A new species of fossil lungfish (Osteichthyes: Dipnoi) from the Cretaceous of Australia. *Journal of Vertebrate Paleontology* **40**, e1822369. doi:10.1080/02724634.2020.1822369

205. Ketchumm HF, Benson RBJ (2011) A new pliosaurid (Sauropterygia, Plesiosauria) from the Oxford Clay Formation (Middle Jurassic, Callovian) of England: evidence for a gracile, longirostrine grade of Early-Middle Jurassic pliosaurids. *Special Papers in Palaeontology* **86**, 109–129.

206. Lang WH, Cookson IC (1935) On a flora, including vascular land plants, associated with *Monograptus*, in rocks of Silurian age, from Victoria, Australia. *Philosophical Transactions of the Royal Society of London. Series B, Biological Sciences* **224**, 421–449.

207. Lautenschlager S, Witzman F, Werneberg I (2016) Palate anatomy and morphofunctional aspects of interpterygoid vacuities in temnospondyl cranial evolution. *Naturwissenschaften* **103**, 79. doi:10.1007/s00114-016-1402-z

208. Leahey LG, Molnar RE, Carpenter K, Witmer LW, Salisbury SW (2015) Cranial osteology of the ankylosaurian dinosaur formerly known as *Minmi* sp. (Ornithischia: Thyreophora) from the Lower Cretaceous Allaru Mudstone of Richmond, Queensland, Australia. *PeerJ* **3**, e1475. doi:10.7717/peerj.1475

209. Lebedev OA, Coates MI (1995) The postcranial skeleton of the Devonian tetrapod *Tulerpeton curtum* Lebedev. *Zoological Journal of the Linnean Society* **114**, 307–348. doi:10.1111/j.1096-3642.1995.tb00119.x

210. Lee MSY, Hutchinson MN, Worthy TH, Archer M, Tennyson AJD, Worthy JP, Scofield RP (2009) Miocene skinks and geckos reveal long-term conservatism of New Zealand's lizard fauna. *Biology Letters* **5**, 833–837. doi:10.1098/rsbl.2009.0440

211. Lee MSY, Jago JB, García-Bellido DC, Edgecombe GD, Gehling JG, Paterson JR (2011) Modern optics in exceptionally preserved eyes of Early Cambrian arthropods from Australia. *Nature* **474**, 631–634. doi:10.1038/nature10097

212. Leu MR (1989) A Late Permian freshwater shark from Eastern Australia. *Palaeontology* **32**, 265–286.

213. Lindley ID (2002) Acanthodian, onychodontid and osteolepidid fish from the middle-upper Taemas Limestone (Early Devonian), Lake Burrinjuck, New South Wales. *Alcheringa* **26**, 103–126. doi:10.1080/03115510208619246

214. Llamas B, Brotherton P, Mitchell KJ, Templeton JEL, Thomson VA, Metcalf JL, Armstrong KN, Kasper M, Richards SM, Camens AB, Lee MSY, Cooper A (2015) Late Pleistocene Australian marsupial DNA clarifies the affinities of extinct megafaunal kangaroos and wallabies. *Molecular Biology and Evolution* **32**, 574–584. doi:10.1093/molbev/msu338

215. Long JA (1985) New information on the head and shoulder girdle of *Canowindra grossi* Thomson, from the Upper Devonian Mandagery Sandstone, New South Wales. *Records of the Australian Museum* **37**, 91–99. doi:10.3853/j.0067-1975.37.1985.338

216. Long JA (1989) A new rhizodontiform fish from the Early Carboniferous of Victoria, Australia, with remarks on the phylogenetic position of the group. *Journal of Vertebrate Paleontology* **9**, 1–17. doi:10.1080/02724634.1989.10011735

217. Long JA (1991) The long history of Australian fossil fishes. In *Vertebrate Palaeontology of Australasia*. (Eds PV Rich, J Monaghan, RF Baird and T Rich) pp. 337–428. Pioneer Design Studios, Melbourne.

218. Long JA (1995) A new plourdosteid arthrodire from the Upper Devonian Gogo Formation of Western Australia. *Palaeontology* **38**, 39–62.

219. Long JA (1997) Ptyctodontid fishes from the Late Devonian Gogo Formation, Western Australia, with a revision of the European genus *Ctenurella* Ørvig 1960. *Geodiversitas* **19**, 515–555.

220. Long JA (1998) *Dinosaurs of Australia and New Zealand and Other Animals of the Mesozoic Era*. UNSW Press, Sydney.

221. Long JA (2003) Middle Devonian to Carboniferous. In *Geology of Victoria*. (Ed. WD Birch) pp. 190–193. Special Publication 23. Geological Society of Victoria, Melbourne.

222. Long JA (2011) *The Rise of Fishes: 500 Million Years of Evolution*. Johns Hopkins University Press, Baltimore.

223. Long J, Archer M, Flannery TF, Hand SJ (2002) *Prehistoric Mammals of Australia and New Guinea*. UNSW Press: Sydney.

224. Long JA, Mark-Kurik E, Johanson Z, Lee MSY, Young GC, Zhu M, Ahlberg PE, Newman M, Jones R, Den Blaauwen J, Choo B, Trinajstic K (2014) Copulation in antiarch placoderms and the origin of gnathostome internal fertilisation. *Nature* **517**, 196–199. doi:10.1038/nature13825

225. Long JA, Trinsjstic K (2010) The Late Devonian Gogo Formation Lägerstatten of Western Australia: exceptional early vertebrate preservation and diversity. *Annual Review of Earth and Planetary Sciences* **38**, 255–279. doi:10.1146/annurev-earth-040809-152416

226. Long JA, Trinajstic KM, Young GC, Senden TJ (2008) Live birth in the Devonian period. *Nature* **453**, 650–652. doi:10.1038/nature06966

227. Long JA, Young GC (1988) Acanthothoracid remains from the Early Devonian of New South Wales, including a complete sclerotic capsule and pelvic girdle. *Memoirs of the Association of Australasian Palaeontologists* **7**, 65–80.

228. Long JA, Young GC, Holland T, Senden TJ, Fitzgerald EMC (2006) An exceptional Devonian fish from Australia sheds light on tetrapod origins. *Nature* **444**, 199–202. doi:10.1038/nature05243

229. Longman HA (1924) A new gigantic marine reptile from the Queensland Cretaceous, *Kronosaurus queenslandicus* new genus and species. *Memoirs of the Queensland Museum* **8**, 26–28.

230. Louys J, Aplin K, Beck RMD, Archer M (2009) Cranial anatomy of Oligo-Miocene koalas (Diprotodontia: Phascolarctidae): stages in the evolution of an extreme leaf-eating specialization. *Journal of Vertebrate Paleontology* **29**, 981–992. doi:10.1671/039.029.0412

231. Louys J, Black K, Archer M, Hand SJ, Godthelp H (2007) Descriptions of koala fossils from the Miocene of Riversleigh, northwestern Queensland and implications for *Litokoala*. *Alcheringa* **31**, 99–110. doi:10.1080/03115510701305082

232. Macken AC, Prideaux GJ, Reed EH (2012) Variation and pattern in the responses of mammal faunas to Late Pleistocene climatic change in southeastern South Australia. *Journal of Quaternary Science* **27**, 415–424. doi:10.1002/jqs.1563

233. Macken AC, Reed EH (2014) Late Quaternary small mammal faunas of the Naracoorte Caves World Heritage Area. *Transactions of the Royal Society of South Australia* **137**, 53–67. doi:10.1080/3721426.2013.10887171

234.. Macphail MK (1996) A habitat for the enigmatic *Wynyardia bassiana* Spencer, 1901. *Alcheringa* **20**, 227–243. doi:10.1080/03115519608619190

235. MacPhee RDE (Ed.) (1999) *Extinctions in Near Time: Causes, Contexts, and Consequences.* Springer, New York.

236. MacPhee RDE (2018) *End of the Megafauna: The Fate of the World's Hugest, Fiercest, and Strangest Animals.* W.W. Norton, New York.

237. Madzia D, Boyd CA, Mazuch M (2017) A basal ornithopod dinosaur from the Cenomanian of the Czech Republic. *Journal of Systematic Palaeontology* **16**, 967–979. doi:10.1080/14772019.2017.1371258

238. Martin MW, Grazhdankin DV, Bowring SA, Evans DA, Fedonkin MA, Kirschvink JL (2000) Age of Neoproterozoic bilatarian body and trace fossils, White Sea, Russia: implications for metazoan evolution. *Science* **288**, 841–845. doi:10.1126/science.288.5467.841

239. Mather EK, Lee MSY, Camens AB, Worthy TH (2021) An exceptional partial skeleton of a new basal raptor (Aves: Accipitridae) from the late Oligocene Namba Formation, South Australia. *Historical Biology.* doi:10.1080/08912963.2021.1966777

240. Matzke-Karasz R, Neil JV, Smith RJ, Symonova R, Morkovsky L, Archer M, Hand SJ, Cloetens P, Tafforeau P (2014) Subcellular preservation in giant ostracod sperm from an early Miocene cave deposit in Australia. *Proceedings. Biological Sciences* **281**, 20140394. doi:10.1098/rspb.2014.0394

241. McHenry CR (2009) Devourer of gods: the palaeoecology of the Cretaceous pliosaur *Kronosaurus queenslandicus.* Unpublished PhD thesis. University of Newcastle, Newcastle.

242. McNamara GC (1990) The Wyandotte local fauna: a new, dated, Pleistocene vertebrate fauna from northern Queensland. *Memoirs of the Queensland Museum* **28**, 285–297.

243. McNamara JA (1994) A new fossil wallaby (Marsupialia: Macropodidae) from the southeast of South Australia. *Records of the South Australian Museum* **27**, 111–115.

244. McSweeney F, Buckeridge J (Eds) (2017) *The Fossils of the Urban Sanctuary: Rickett's Point Victoria 3193.* Greypath Productions, Melbourne.

245. Mead JL, Steadman DW, Bedford SH, Bell CJ, Spriggs M (2002) New extinct mekosuchine crocodile from Vanuatu, South Pacific. *Copeia* (3), 632–641. doi:10.1643/0045-8511(2002)002[0632:NEMCFV]2.0.CO;2

246. Megirian D, Murray PF, Willis P (1991) A new crocodile of the gavial ecomorph morphology from the Miocene of northern Australia. *The Beagle: Records of the Museums and Art Galleries of the Northern Territory* **8**, 135–157. doi:10.5962/p.262817

247. Merrilees D (1968) Man the destroyer: Late Quaternary changes in the Australian marsupial fauna. *Journal and Proceedings of the Royal Society of Western Australia* **57**, 1–24.

248. Merrilees D (1968) South-western Australian occurrences of *Sthenurus* (Marsupialia, Macropodidae), including *Sthenurus brownei* sp. nov. *Journal and Proceedings of the Royal Society of Western Australia* **50**, 65–79.

249. Miles RS (1971) The Holonematidae (placoderm fishes): a review based on new specimens of *Holonema* from the Upper Devonian of Western Australia. *Philosophical Transactions of the Royal Society of London* **263**(849), 101–234.

250. Millener PR (1988) Contributions to New Zealand's Late Quaternary avifauna. I: *Pachyplichas*, a new genus of wren (Aves: Acanthisittidae), with two new species. *Journal of the Royal Society of New Zealand* **18**, 383–406. doi:10.1080/03036758.1988.10426464

251. Millener PR (1996) Extinct birds. In *The Chatham Islands: Heritage and Conservation*, pp. 113–120. Canterbury University Press in association with New Zealand Department of Conservation, Christchurch.

252. Millener PR (1999) The history of the Chatham Islands' bird fauna of the last 7000 years: a chronicle of change and extinction. In *Proceedings of the 4th International meeting of the Society of Avian Paleontology and Evolution*, Washington DC, June 1996. *Smithsonian Contributions to Paleobiology* 89, 85–109.

253. Miller AH (1966) The fossil pelicans of Australia. *Memoirs of the Queensland Museum* **14**, 181–190.

254. Mitchell KJ, Wood JR, Scofield RP, Llamas B, Cooper A (2014) Ancient mitochondrial genome reveals unsuspected taxonomic affinity of the extinct Chatham duck (*Pachyanas chathamica*) and resolves divergence times for New Zealand and sub-Antarctic brown teals. *Molecular Phylogenetics and Evolution* **70**, 420–428. doi:10.1016/j.ympev.2013.08.017

255. Molnar RE (1986) An enantiornithine bird from the Lower Cretaceous of Queensland, Australia. *Nature* **322**, 736–738 doi:10.1038/322736a0.

256. Molnar RE (1996) Observations on the Australian ornithopod dinosaur *Muttaburrasaurus*. *Memoirs of the Queensland Museum* **39**, 639–652.

257. Molnar RE (2001) Armor of the small ankylosaur *Minmi*. In *The Armored Dinosaurs*. (Ed. K Carpenter) pp. 341–362. Indiana University Press, Bloomington.

258. Molnar RE (2004) *Dragons in the Dust: The Paleobiology of the Giant Monitor Lizard Megalania*. Indiana University Press, Bloomington.

259. Molnar RE, Flannery TF, Rich THV (1981) An allosaurid theropod dinosaur from the Early Cretaceous of Victoria, Australia. *Alcheringa* **5**, 141–146. doi:10.1080/03115518108565427

260. Molnar RE, Thulborn RA (2008) An incomplete pterosaur skull from the Cretaceous of north-central Queensland, Australia. *Arquivos do Museu Nacional, Rio de Janeiro* **65**, 461–470.

261. Molnar RE, Worthy TH, Willis PMA (2002) An extinct Pleistocene endemic mekosuchine crocodylian from Fiji. *Journal of Vertebrate Paleontology* **22**, 612–628. doi:10.1671/0272-4634(2002)022[0612:AEPEMC]2.0.CO;2

262. Montanari S, Louys J, Price GJ (2013) Pliocene paleoenvironments of southeastern Queensland, Australia inferred from stable isotopes of marsupial tooth enamel. *PLoS One* **8**, e66221. doi:10.1371/journal.pone.0066221

263. Moriarty KC, McCulloch MT, Wells RT, McDowell MC (2000) Mid-Pleistocene cave fills, megafaunal remains and climate change at Naracoorte, South Australia: towards a predictive model using U-Th dating of speleothems. *Palaeogeography, Palaeoclimatology, Palaeoecology* **159**, 113–143. doi:10.1016/S0031-0182(00)00036-5

264. Mountain M-J (1991) Highland New Guinea hunter-gatherers: the evidence of Nombe Rockshelter, Simbu with emphasis on the Pleistocene. Unpublished PhD thesis. Australian National University, Canberra.

265. Mourer-Chauviré C, Balouet JC (1992) Description of the skull of the genus *Sylviornis* Poplin, 1980 (Aves: Galliformes, Sylviornithidae new family), a giant extinct bird from the Holocene of New Caledonia. In *Proceedings of the International Symposium on Insular Vertebrate Evolution: The Palaeontological Approach. Monografies de la Societat d'Història Natural de les Balears* **12**, 205–218.

266. Muirhead J, Archer M (1990) *Nimbacinus dicksoni*, a plesiomorphic thylacine (Marsupialia: Thylacinidae) from Tertiary deposits of Queensland and the Northern Territory. *Memoirs of the Queensland Museum* **28**, 203–221.

267. Muirhead J, Filan S (1995) *Yarala burchfieldi*, a plesiomorphic bandicoot (Marsupialia, Peramelemorphia) from Oligo-Miocene deposits of Riversleigh, northwestern Queensland. *Journal of Paleontology* **69**, 127–134. doi:10.1017/S0022336000026986

268. Muirhead J, Wroe S (1998) A new genus and species, *Badjcinus turnbulli* gen. et sp. nov. (Thylacinidae: Marsupialia), from the late

Oligocene of Riversleigh, northern Australia, and an investigation of thylacinid phylogeny. *Journal of Vertebrate Paleontology* **18**, 612–626. doi:10.1080/02724634.1998.10011088

269. Murray P (1986) *Propalorchestes novaculacephalus* gen. et sp. nov., a new palorchestid (Diprotodontoidea: Marsupialia) from the Middle Miocene Camfield Beds, Northern Territory, Australia. *The Beagle: Records of the Museums and Art Galleries of the Northern Territory* **3**, 195–211.

270. Murray P (1991) The Pleistocene megafauna of Australia. In *Vertebrate Palaeontology of Australasia.* (Eds P Vickers-Rich, JM Monaghan, RF Baird and TH Rich) pp. 1071–1164. Pioneer Design Studio, Melbourne.

271. Murray PF (1992) The smallest New Guinea zygomaturines derived dwarfs or relict plesiomorphs? *The Beagle: Records of the Museums and Art Galleries of the Northern Territory* **9**, 89–110. doi:10.5962/p.263120

272. Murray PF (1992) Thinheads, thickheads and airheads: functional craniology of some diprotodontian marsupials. *The Beagle: Records of the Museums and Art Galleries of the Northern Territory* **9**, 71–87. doi:10.5962/p.263119

273. Murray P, Megirian D (1992) Continuity and contrast in middle and late Miocene vertebrate communities from the Northern Territory. *The Beagle: Records of the Museums and Art Galleries of the Northern Territory* **9**, 195–217. doi:10.5962/p.263125

274. Murray P, Megirian D, Rich T, Plane M, Black K, Archer M, Hand SJ, Vickers-Rich P (2000) *Morphology, systematics and evolution of the marsupial genus* Neohelos *Stirton (Diprotodontidae, Zygomaturinae).* Research Report No. 6. Museums and Art Galleries of the Northern Territory, Darwin.

275. Musser AM, Archer M (1998) New information about the skull and dentary of the Miocene platypus *Obdurodon dicksoni* and a discussion of ornithorhynchid relationships. *Philosophical Transactions of the Royal Society of London. Series B, Biological Sciences* **353**, 1063–1079. doi:10.1098/rstb.1998.0266

276. Myers T, Archer M (1997) *Kuterintja ngama* (Marsupialia, Ilariidae): a revised systematic analysis based on material from the late Oligocene of Riversleigh, northwestern Queensland. *Memoirs of the Queensland Museum* **41**, 379–392.

277. Myers T, Archer M, Krikmann A, Pledge N (1999) Diversity and evolutionary relationships of ilariids, wynyardiids, vombatids and related groups of marsupials. *Australian Mammalogy* **21**, 18–19, 34–45.

278. Myers TJ, Black KH, Archer M, Hand SJ (2017) The identification of Oligo-Miocene mammalian palaeocommunities from the Riversleigh World Heritage Area, Australia and an appraisal of palaeoecological techniques. *PeerJ* **5**, e3511. doi:10.7717/peerj.3511

279. Nguyen JM, Boles WE, Hand SJ (2010) New material of *Barawertornis tedfordi,* a dromornithid bird from the Oligo-Miocene of Australia, and its phylogenetic implications. *Records of the Australian Museum* **62**, 45–60. doi:10.3853/j.0067-1975.62.2010.1539

280. Nutman AP, Bennett VC, Friend CRL, van Kranendonk MJ, Chivas A (2016) Rapid emergence of life shown by discovery of 3,700-million-year-old microbial structures. *Nature* **537**, 535–538. doi:10.1038/nature19355

281. Owen R (1838) In *Three Expeditions into the Interior of Eastern Australia, with Descriptions of the Recently Explored Region of Australia Felix, and of the Present Colony of New South Wales.* (Ed. TL Mitchell) vol. 2, pp. 362–363. T & W Boone, London.

282. Owen R (1866) On *Dinornis* (Part X): containing a description of part of the skeleton of a flightless bird indicative of a new genus and species (*Cnemiornis calcitrans,* Ow.). *Transactions of the Zoological Society of London V* **5**(5), 395–404 [plates LXIII–LXVII.]. doi:10.1111/j.1096-3642.1866.tb00650.x

283. Owen R (1871) On *Dinornis* (Part XV): containing a description of the skull, femur, tibia, fibula, and metatarsus of *Aptornis defossor,* Owen, from near Oamaru, Middle Island, New Zealand; with additional observations on *Aptornis otidiformis,*

on *Notornis mantelli*, and on *Dinornis curtus*. *Transactions of the Zoological Society, London* **7**(5), 353–380 [plates 40-46].

284. Owen R (1887) Description of fossil remains of two species of a Megalanian genus (*Meiolania*) from "Lord Howe's Island". *Philosophical Transactions of the Royal Society of London* **1886**, 471–480.

285. Palci A, Hutchinson MN, Caldwell MW, Scanlon JD, Lee MSY (2018) Palaeoecological inferences for the fossil Australian snakes *Yurlunggur* and *Wonambi* (Serpentes, Madtsoiidae). *Royal Society Open Science* **5**, 172012. doi:10.1098/rsos.172012

286. Park T, Fitzgerald EMG (2012) A review of Australian fossil penguins (Aves: Sphenisciformes). *Memoirs of the Museum of Victoria* **69**, 309–325. doi:10.24199/j.mmv.2012.69.06

287. Paterson JR, Edgecombe GD, García-Bellido DC, Jago JB, Gehling JG (2010) Nektaspid arthropods from the lower Cambrian Emu Bay Shale Lagerstätte, South Australia, with a reassessment of lamellipedian relationships. *Palaeontology* **53**, 377–402. doi:10.1111/j.1475-4983.2010.00932.x

288. Paterson JR, Edgecombe GD, Jago JB (2015) The 'great appendage' arthropod *Tanglangia*: biogeographic connections between early Cambrian biotas of Australia and South China. *Gondwana Research* **27**, 1667–1672. doi:10.1016/j.gr.2014.02.008

289. Paterson JR, Garcia-Bellido DC, Jago JB, Gehling JG, Lee MSY, Edgecombe GD (2016) The Emu Bay Shale Konservat-Lagerstatte: a view of Cambrian life from East Gondwana. *Journal of the Geological Society* **173**, 1–11. doi:10.1144/jgs2015-083

290. Paterson JR, García-Bellido DC, Lee MSY, Brock GA, Jago JB, Edgecombe GD (2011) Acute vision in the giant Cambrian predator *Anomalocaris* and the origin of compound eyes. *Nature* **480**, 237–240. doi:10.1038/nature10689

291. Pentland AH, Poropat SF (2019) Reappraisal of *Mythunga camara* Molnar & Thulborn, 2007 (Pterosauria, Pterodactyloidea, Anhangueria) from the upper Albian Toolebuc Formation of Queensland, Australia. *Cretaceous Research* **93**, 151–169. doi:10.1016/j.cretres.2018.09.011

292. Pian R, Archer M, Hand SJ, Beck RMD, Cody A (2016) The upper dentition and relationships of the enigmatic Australian Cretaceous mammal *Kollikodon ritchiei*. *Memoirs of the Museum of Victoria* **74**, 97–105. doi:10.24199/j.mmv.2016.74.10

293. Pinheiro FL, Franca MAG, Lacerda MB, Butler RJ, Schultz CL (2016) An exceptional fossil skull from South America and the origins of the archosauriform radiation. *Scientific Reports* **6**, 22817. doi:10.1038/srep22817

294. Plane MD (1967) Stratigraphy and vertebrate fauna of the Otibanda Formation, New Guinea. *Bulletin of the Bureau of Mineral Resources, Geology and Geophysics* **86**, 1–64.

295. Plane MD (1976) The occurrence of *Thylacinus* in Tertiary rocks from Papua New Guinea. *Journal of Australian Geology and Geophysics* **I**, 78–79.

296. Pledge NS (1981) The giant rat-kangaroo *Propleopus oscillans* (De Vis), (Potoroidae: Marsupialia) in South Australia. *Transactions of the Royal Society of South Australia* **105**, 41–47.

297. Pledge NS (1982) Enigmatic *Ektopodon*: a case history of palaeontological interpretation. In *The Fossil Vertebrate Record of Australasia*. (Eds PV Rich and EM Thompson) pp. 479–488. Monash University Publications Committee, Melbourne.

298. Pledge NS (1991) Reconstructing the natural history of extinct animals: *Ektopodon* as a case history. In *Vertebrate Palaeontology of Australasia*. (Eds P Vickers-Rich, JM Monaghan, RF Baird and TH Rich) pp. 247–266. Pioneer Design Studio, Melbourne.

299. Pledge NS (1994) Fossils of the lake: a history of the Lake Callabonna excavations. *Records of the South Australian Museum* **27**, 65–77.

300. Pledge NS (2005) The Riversleigh wynyardiids. *Memoirs of the Queensland Museum* **51**, 135–169.

301. Pledge NS (2010) A new koala (Marsupialia: Phascolarctidae) from the late Oligocene Etadunna Formation, Lake Eyre Basin, South Australia. *Australian Mammalogy* **32**, 79–86. doi:10.1071/AM09014

302. Pledge NS (2016) New specimens of ektopodontids (Marsupialia: Ektopodontidae) from South Australia. *Memoirs of the Museum of Victoria* **74**, 173–187. doi:10.24199/j.mmv.2016.74.15

303. Pledge NS, Archer M, Hand SJ, Godthelp H (1999) Additions to knowledge about ektopodontids (Marsupialia: Ektopodontidae): including a new species *Ektopodon litolophus*. *Records of the Western Australian Museum* **57**(Supplement), 255–264.

304. Poplin F, Mourer-Chauvire C (1985) *Sylviornis neocaledoniae* (Aves, Galliformes, Megapodiidae), *oiseau geant eteint de L'ile des Pins (Nouvelle-Caledonie)*. *Geobios* **18**, 73–105. doi:10.1016/S0016-6995(85)80182-0

305. Poropat SF, Kool L, Vickers-Rich P, Rich TH (2017) Oldest meiolaniid turtle remains from Australia: evidence from the Eocene Kerosene Creek Member of the Rundle Formation, Queensland. *Alcheringa* **41**, 231–239. doi:10.1080/03115518.2016.1224441

306. Poropat SF, Mannion PD, Upchurch P, Hocknull SA, Kear BP, Kundrat M, Tischler TR, Sloan T, Sinapius GHK, Elliott JA, Elliott DA (2016) New Australian sauropods shed light on Cretaceous dinosaur palaeobiogeography. *Scientific Reports* **6**, 34467. doi:10.1038/srep34467

307. Poropat SF, Nair JP, Syme CE, Mannion PD, Upchurch P, Hocknull SA, Cook AG, Tischler TR, Holland T (2017) Reappraisal of *Austrosaurus mckillopi* Longman, 1933 from the Allaru Mudstone of Queensland, Australia's first named Cretaceous sauropod dinosaur. *Alcheringa* **41**, 543–580. doi:10.1080/03115518.2017.1334826

308. Poropat SF, Upchurch P, Mannion PD, Hocknull SA, Kear BP, Sloan T, Sinapius GHK, Elliott DA (2015) Revision of the sauropod dinosaur *Diamantinasaurus matildae* Hocknull et al., (2009) from the mid-Cretaceous of Australia: implications for Gondwanan titanosauriform dispersal. *Gondwana Research* **27**, 995–1033. doi:10.1016/j.gr.2014.03.014

309. Pregill GK, Dye T (1989) Prehistoric extinction of giant iguanas in Tonga. *Copeia* **1989**, 505–508. doi:10.2307/1445455

310. Pregill GK, Steadman DW (2004) South Pacific iguanas: human impacts and a new species. *Journal of Herpetology* **38**, 15–21. doi:10.1670/73-03A

311. Price G (2008) Taxonomy and palaeobiology of the largest-ever marsupial *Diprotodon* Owen, 1838 (Diprotodontidae, Marsupialia). *Zoological Journal of the Linnean Society* **153**, 369–397. doi:10.1111/j.1096-3642.2008.00387.x

312. Price GJ (2013) Quaternary. In *Geology of Queensland*. (Ed. PA Jell) pp. 653–685. Geological Survey of Queensland, Brisbane.

313. Price GJ, Ferguson KJ, Webb GE, Feng Y, Higgins P, Nguyen AD, Zhao J, Joannes-Boyau R, Louys J (2017) Seasonal migration of marsupial megafauna in Pleistocene Sahul (Australia–New Guinea). *Proceedings. Biological Sciences* **284**, 20170785. doi:10.1098/rspb.2017.0785

314. Price GJ, Louys J, Cramb J, Feng Y-X, Zhao J-X, Hocknull SA, Webb GE, Nguyen AD, Joannes-Boyau R (2015) Temporal overlap of humans and giant lizards (Varanidae; Squamata) in Pleistocene Australia. *Quaternary Science Reviews* **125**, 98–105. doi:10.1016/j.quascirev.2015.08.013

315. Price GJ, Sobbe IH (2005) Pleistocene palaeoecology and environmental change on the Darling Downs, southeastern Queensland, Australia. *Memoirs of the Queensland Museum* **51**, 171–201.

316. Price GJ, Webb GE, Zhao J, Feng Y-X, Murray AS, Cooke BN, Hocknull SA, Sobbe IH (2011) Dating megafaunal extinction on the Pleistocene Darling Downs, eastern Australia: the promise and pitfalls of dating as a test of extinction hypotheses. *Quaternary Science Reviews* **30**, 899–914. doi:10.1016/j.quascirev.2011.01.011

317. Price GJ, Zhao J-X, Feng Y-X, Hocknull SA (2009) New U/Th ages for Pleistocene megafauna deposits of southeastern Queensland, Australia. *Journal of Asian Earth Sciences* **34**, 190–197. doi:10.1016/j.jseaes.2008.04.008

318. Prideaux G (2004) *Systematics and Evolution of Sthenurine Kangaroos*. UC Publications in Geological Sciences, Berkeley.

319. Prideaux GJ, Long JA, Ayliffe LK, Hellstrom JC, Pillans B, Bowles WE, Hutchinson MN, Roberts RG, Cupper ML, Arnold LJ, Devine PD, Warburton NM (2007) An arid-adapted middle Pleistocene vertebrate fauna from south-central Australia. *Nature* **445**, 422–425. doi:10.1038/nature05471

320. Prideaux GJ, Warburton NM (2008) A new Pleistocene tree-kangaroo (Diprotodontia: Macropodidae) from the Nullarbor Plain of south-central Australia. *Journal of Vertebrate Paleontology* **28**, 463–478. doi:10.1671/0272-4634(2008)28[463:ANPTDM]2.0.CO;2

321. Rampino MR, Caldeira K, Zhu Y (2021) A pulse of the Earth: a 27.5-Myr underlying cycle in coordinated geological events over the last 260 Myr. *Geoscience Frontiers* **12**(6), 101245. doi:10.1016/j.gsf.2021.101245

322. Rawlence NJ, Metcalf JL, Wood JR, Worthy TH, Austin JJ, Cooper A (2012) The effect of climate and environmental change on the megafaunal moa of New Zealand in the absence of humans. *Quaternary Science Reviews* **50**, 141–153. doi:10.1016/j.quascirev.2012.07.004

323. Retallack GJ (2013) Ediacaran life on land. *Nature* **493**, 89–92. doi:10.1038/nature11777

324. Rich PV (1979) The Dromornithidae, an extinct family of large ground birds endemic to Australia. *Bulletin of the Bureau of Mineral Resources, Geology and Geophysics* **184**, 1–194.

325. Rich PV, van Tets G (1981) The fossil pelicans of Australia. *Records of the South Australian Museum* **18**, 235–264.

326. Rich PV, van Tets GF, Knight F (Eds) (1985) *Kadimakara: Extinct Vertebrates of Australia.* Pioneer Design Studio, Melbourne.

327. Rich TH, Lawson PF, Vickers-Rich P, Tedford RH (2019) R.A. Stirton: pioneer of Australian mammalian palaeontology. *Transactions of the Royal Society of South Australia* **143**, 244–282. doi:10.1080/03721426.2019.1602244

328. Rich TH, Vickers-Rich P (1989) Polar dinosaurs and biotas of the Early Cretaceous of southeastern Australia. *National Geographic Research* **5**, 15–53.

329. Rich TH, Vickers-Rich P (1994) Neoceratopsians and ornithomimosaurs: dinosaurs of Gondwana origin? *National Geographic Research and Exploration* **10**, 129–131.

330. Rich TH, Vickers-Rich P (2002) *Dinosaurs of Darkness.* Indiana University Press, Bloomington.

331. Richards HL, Wells RT, Evans AR, Fitzgerald EMG, Adams JW (2019) The extraordinary osteology and functional morphology of the limbs in Palorchestidae, a family of strange extinct marsupial giants. *PLoS One* **14**(9), e0221824. doi:10.1371/journal.pone.0221824

332. Ride WDL, Pridmore PA, Barwick RE (1997) Towards a biology of *Propleopus oscillans* (Marsupialia: Propleopinae, Hypsiprymnodontidae). *Proceedings of the Linnean Society of New South Wales* **117**, 243–328.

333. Rieppel O, Kluge AG, Zaher H (2003) Testing the phylogenetic relationships of the Pleistocene snake *Wonambi naracoortensis* Smith. *Journal of Vertebrate Paleontology* **22**, 812–829. doi:10.1671/0272-4634(2002)022[0812:TTPROT]2.0.CO;2

334. Ristevski J, Yates AM, Price GJ, Molnar RE, Weisbecker V, Salisbury SW (2020) Australia's prehistoric 'swamp king': revision of the Plio-Pleistocene crocodylian genus *Pallimnarchus* de Vis, 1886. *PeerJ* **8**, e10466. doi:10.7717/peerj.10466

335. Ritchie A (1973) *Wuttagoonaspis* gen. nov., an unusual arthrodire from the Devonian of western New South Wales, Australia. *Palaeontographica* **143A**, 58–72.

336. Ritchie A (2004) A new genus and species of groenlandaspidid artrhrodire (Pisces: Placodermi) from the early-Middle Devonian Mulga Downs Group of western New South Wales, Australia. *Fossils and Strata* **50**, 56–81.

337. Ritchie A (2006) The great Devonian fish kill at Canowindra. In *Evolution and Biogeography of Australasian Vertebrates.* (Eds JR Merrick, M Archer, GM Hickey and MSY Lee) pp. 159–184. Auscipub, Sydney.

338. Ritchie A, Gilbert Tomlinson J (1977) First Ordovician vertebrates form the Southern Hemisphere. *Alcheringa* **1**, 351–368. doi:10.1080/03115517708527770

339. Roberts KK, Archer M, Hand SJ, Godthelp H (2009) New Australian Oligocene to Miocene ringtail possums (Pseudocheiridae) and revision of the genus *Marlu. Palaeontology* **52**, 441–456. doi:10.1111/j.1475-4983.2009.00852.x

340. Romer AS, Lewis AD (1959) A mounted skeleton of the giant plesiosaur *Kronosaurus. Breviora* **112**, 1–15.

341. Rovinsky DS, Evans AR, Adams JW (2019) The pre-Pleistocene fossil thylacinids (Dasyuromorphia: Thylacinidae) and the evolutionary context of the modern thylacine. *PeerJ* **7**, e7457. doi:10.7717/peerj.7457

342. Sachs S (2005) *Tuarangisaurus australis* sp. nov. (Plesiosauria: Elasmosauridae) from the Lower Cretaceous of northeastern Queensland, with additional notes on the phylogeny of the Elasmosauridae. *Memoirs of the Queensland Museum* **50**, 425–440.

343. Salisbury SW, Molnar R, Frey E, Willis PW (2006) The origin of modern crocodyliforms: new evidence from the Cretaceous of Australia. *Proceedings. Biological Sciences* **273**, 2439–2448. doi:10.1098/rspb.2006.3613

344. Salisbury SW, Romilio A, Herne MC, Tucker RT, Nair JP (2016) The dinosaurian ichnofauna of the Lower Cretaceous (Valanginian–Barremian) Broome Sandstone of the Walmadany Area (James Price Point), Dampier Peninsula, Western Australia. *Journal of Vertebrate Paleontology* **36**(supp.), 1–152. doi:10.1080/02724634.2016.1269539

345. Salisbury SW, Willis PMA (1996) A new crocodylian from the early Eocene of southeastern Queensland and a preliminary investigation into the phylogenetic relationships of crocodyloids. *Alcheringa* **20**, 179–226. doi:10.1080/03115519608619189

346. Sansom IJ, Davies NS, Coates MI, Nicoll RS, Ritchie A (2012) Chondrichthyan-like scales from the Middle Ordovician of Australia. *Palaeontology* **55**, 243–247. doi:10.1111/j.1475-4983.2012.01127.x

347. Scanlon JD (2005) Australia's oldest known snakes: *Patagoniophis, Alamitophis*, and cf. *Madtsoia* (Squamata: Madtsoiidae) from the Eocene of Queensland. *Memoirs of the Queensland Museum* **51**, 215–235.

348. Scanlon JD (2005) Cranial morphology of the Plio-Pleistocene giant madtsoiid snake *Wonambi naracoortensis*. *Acta Palaeontologica Polonica* **50**, 139–180.

349. Schwartz LRS (2016) A revised faunal list and geological setting for Bullock Creek, a Camfieldian site from the Northern Territory of Australia. *Memoirs of the Museum of Victoria* **74**, 263–290. doi:10.24199/j.mmv.2016.74.20

350. Scofield PR, Ashwell KWS (2009) Rapid somatic expansion causes the brain to lag behind: the case of the brain and behaviour of New Zealand's Haast's Eagle (*Harpagornis moorei*). *Journal of Vertebrate Paleontology* **29**, 637–649. doi:10.1671/039.029.0325

351. Scofield RP, Worthy TH, Tennyson AJD (2010) A heron (Aves: Ardeidae) from the Early Miocene St Bathans fauna of southern New Zealand. *Records of the Australian Museum* **62**, 89–104. doi:10.3853/j.0067-1975.62.2010.1542

352. Sferco E, Lopez-Arbarello A, Baez AM (2015) Phylogenetic relationships of *Luisiella feruglioi* (Bordas) and the recognition of a new clade of freshwater teleosts from the Jurassic of Gondwana. *BMC Evolutionary Biology* **15**, 268. doi:10.1186/s12862-015-0551-6

353. Sharp AC (2016) A quantitative comparative analysis of the size of the frontoparietal sinuses and brain in vombatiform marsupials. *Memoirs of the Museum of Victoria* **74**, 331–342. doi:10.24199/j.mmv.2016.74.23

354. Shea GM, Hutchinson MN (1992) A new species of lizard (*Tiliqua*) from the Miocene of Riversleigh, Queensland. *Memoirs of the Queensland Museum* **32**, 303–310.

355. Shu D, Luo H-L, Conway Morris S, Zhang X, Hu S-X, Chen L, Han J, Zhu M, Li Y, Chen L-Z (1999) Lower Cambrian vertebrates from South China. *Nature* **402**, 42–46. doi:10.1038/46965

356. Shubin NH, Daeschler EB, Jenkins FA Jr (2006) The pectoral fin of *Tiktaalik* and the origin of the tetrapod limb. *Nature* **440**, 764–771. doi:10.1038/nature04637

357. Sigé B, Archer M, Crochet J-Y, Godthelp H, Hand SJ, Beck RMD (2009) *Chulpasia* and *Thylacotinga*, late Paleocene-earliest Eocene trans-Antarctic Gondwanan bunodont marsupials: new data from Australia. *Geobios* **42**, 813–823. doi:10.1016/j.geobios.2009.08.001

358. Simpson GG (1970) Miocene penguins from Victoria, Australia, and Chubut, Argentina. *Memoirs of the National Museum of Victoria* **31**, 17–23. doi:10.24199/j.mmv.1970.31.02

359. Smith MM, Hall BK (1990) Development and evolutionary origins of vertebrate skeletogenic and odontogenic tissues. *Biological Reviews of the Cambridge Philosophical Society* **65**, 277–373. doi:10.1111/j.1469-185X.1990.tb01427.x

360. Smith N, Pol D (2007) Anatomy of a basal sauropodomorph dinosaur from the Early Jurassic Hanson Formation of Antarctica. *Acta Palaeontologica Polonica* **52**, 657–674.

361. Smith ND, Makovicky PJ, Agnolin FL, Ezcurra MD, Pais DF, Salisbury SW (2008) A megaraptor-like theropod (Dinosauria: Tetanurae) in Australia: support for faunal exchange across eastern and western Gondwana in the mid-Cretaceous. *Proceedings. Biological Sciences* **275**, 2085–2093. doi:10.1098/rspb.2008.0504

362. Smith ND, Makovicky PJ, Hammer WR, Currie PJ (2007) Osteology of *Cryolophosaurus ellioti* (Dinosauria: Theropoda) from the Early Jurassic of Antarctica and implications for early theropod evolution. *Zoological Journal of the Linnean Society* **151**, 377–421. doi:10.1111/j.1096-3642.2007.00325.x

363. Smith ND, Makovicky PJ, Pol D, Hammer WR, Currie PJ (2007) *The Dinosaurs of the Early Jurassic Hanson Formation of the Central Transantarctic Mountains: Phylogenetic Review and Synthesis*. US Geological Survey and the National Academies, Washington, DC.

364. Sperling EA, Vinther J (2010) A placozoan affinity for *Dickinsonia* and the evolution of late Proterozoic metazoan feeding modes. *Evolution & Development* **12**, 201–209. doi:10.1111/j.1525-142X.2010.00404.x

365. Steadman DW (1989) New species and records of birds (Aves: Megapodiidae, Columbidae) from an archeological site on Lifuka, Tonga. *Proceedings of the Biological Society of Washington* **102**, 537–552.

366. Steadman DW, Pregill GK, Burley DV (2002) Rapid prehistoric extinction of iguanas and birds in Polynesia. *Proceedings of the National Academy of Sciences of the United States of America* **99**, 3673–3677. doi:10.1073/pnas.072079299

367. Steadman DW, Takano OM (2020) A new genus and species of pigeon (Aves, Columbidae) from the Kingdom of Tonga, with an evaluation of hindlimb osteology of columbids from Oceania. *Zootaxa* **4810**(3), 401–420.

368. Stein M, Hand SJ, Archer M (2016) A new crocodile displaying extreme constriction of the mandible, from the Late Oligocene of Riversleigh, Australia. *Journal of Vertebrate Paleontology* **36**, e1179041. doi:10.1080/02724634.2016.1179041

369. Stein M, Hand SJ, Archer M, Wroe S, Wilson LAB (2020) Quantitatively assessing mekosuchine crocodile locomotion by geometric morphometric and finite element analysis of the forelimb. *PeerJ* **8**, e9349. doi:10.7717/peerj.9349

370. Stein M, Salisbury S, Hand SJ, Archer M, Godthelp H (2012) Humeral morphology of the early Eocene mekosuchine crocodile *Kambara* from the Tingamarra Local Fauna southeastern Queensland, Australia. *Alcheringa* **36**, 473–486. doi:10.1080/03115518.2012.671697

371. Sterli J (2015) A review of the fossil record of Gondwanan turtles of the clade Meiolaniformes. *Bulletin - Peabody Museum of Natural History* **56**, 21–45. doi:10.3374/014.056.0102

372. Stirling EC, Zietz AHC (1898) Preliminary notes on *Phascolonus gigas*, Owen, and its identity with *Sceparnodon ramsayi*, Owen. *Transactions of the Royal Society of South Australia* **23**, 123–133.

373. Stirton R (1967) A diprotodontid from the Miocene Kutjamarpu Fauna, South Australia. *Bulletin of the Bureau of Mineral Resources, Geology and Geophysics* **85**, 45–51.

374. Stirton RA, Tedford RH, Woodburne MO (1967) A new Tertiary formation and fauna from the Tirari Desert, South Australia. *Records of the South Australian Museum* **15**, 427–461.

375. Sutton A, Mountain M-J, Aplin K, Bulmer S, Denham T (2009) Archaeozoological records for the Highlands of New Guinea: a review of current evidence. *Australian Archaeology* **69**, 41–58. doi:10.1080/03122417.2009.11681900

376. Tanaka Y, Fordyce RE (2014) Fossil dolphin *Otekaikea marplesi* (Latest Oligocene, New Zealand) expands the morphological and taxonomic diversity of Oligocene cetaceans. *PLoS One* **9**(9), e107972. doi:10.1371/journal.pone.0107972

377. Tanaka Y, Fordyce RE (2015) A new Oligo-Miocene dolphin from New Zealand: *Otekaikea huata* expands diversity of the early Platanistoidea. *Palaeontologia Electronica* **18.2.23A**, 1–71. doi:10.26879/518

378. Tanaka Y, Fordyce RE (2015) Historically significant late Oligocene dolphin *Microcetus hectori* Benham 1935: a new species of *Waipatia* (Platanistoidea). *Journal of the Royal Society of New Zealand* **45**(3), 135–150. doi:10.1080/03036758.2015.1016046

379. Tanaka Y, Fordyce RE (2017) *Awamokoa tokarahi*, a new basal dolphin in the Platanistoidea (late Oligocene, New Zealand). *Journal of Systematic Palaeontology* **15**(5), 365–386. doi:10.1080/14772019.2016.1202339

380. Tarhan LG, Droser ML, Gehling JG (2015) Depositional and preservational environments of the Ediacara Member, Rawnsley Quartzite (South Australia): assessment of paleoenvironmental proxies and the timing of 'ferruginization'. *Palaeogeography, Palaeoclimatology, Palaeoecology* **434**, 4–13. doi:10.1016/j.palaeo.2015.04.026

381. Tedford RH (1973) The Diprotodons of Lake Callabonna. *Australian Natural History* **17**, 349–354.

382. Tedford RH, Woodburne MO (1987) The Ilariidae, a new family of vombatiform marsupials from Miocene strata of South Australia and an evaluation of the homology of molar cusps in the Diprotodontia. In *Possums and Opossums: Studies in Evolution*. (Ed. M Archer) pp. 401–418. Surrey Beatty & Sons, Sydney.

383. Teichert C, Glenister BF (1954) Early Ordovician cephalopod fauna from northwestern Australia. *Bulletins of American Paleontology* **35**(150), 153–258.

384. Tennyson AJD, Martinson P (2007) *Extinct Birds of New Zealand*, revised edn. Te Papa Press, Wellington.

385. Ter P (2016) Ichnology and palaeoecology of the Neogene Beaumaris Sandstone: a reconstruction of palaeoenvironments using trace fossils as interpretive tools. MSc by Research, RMIT University, Melbourne.

386. Thomson KS, Campbell KSW (1971) The structure and relationship of the primitive Devonian lungfish *Dipnorhynchus sussmilchi* (Ethridge). *Bulletin - Peabody Museum of Natural History* **38**, 1–109.

387. Thorn KM, Hutchinson MN, Archer M, Lee MSY (2019) A new scincid lizard from the Miocene of northern Australia, and the evolutionary history of social skinks (Scincidae: Egerniinae). *Journal of Vertebrate Paleontology* **39**, e1577873. doi:10.1080/02724634.2019.1577873

388. Thorn KM, Hutchinson MH, Lee MSY, Brown N, Camens AB, Worthy TH (2021) A new species of *Proegernia* from the Namba Formation in South Australia and the early evolution and environment of Australian egerniine skinks. *Royal Society Open Science* **8**, 201686. doi:10.1098/rsos.201686

389. Thulborn RA (1983) A mammal-like reptile from Australia. *Nature* **303**, 330–331. doi:10.1038/303330a0

390. Thulborn RA (1986) The Australian Triassic reptile *Tasmaniosaurus triassicus* (Thecodontia: Proterosuchia). *Journal of Vertebrate Paleontology* **6**, 123–142. doi:10.1080/02724634.1986.10011606

391. Thulborn RA, Hamley T, Foulkes P (1994) Preliminary report on sauropod dinosaur tracks in the Broome Sandstone (lower Cretaceous) of Western Australia. *Gaia* **10**, 85–94.

392. Thulborn T, Turner S (2003) The last dicynodont: an Australian Cretaceous relict. *Proceedings. Biological Sciences* **270**, 985–993. doi:10.1098/rspb.2002.2296

393. Travouillon KJ, Beck RMD, Hand SJ, Archer M (2013) The oldest fossil record of bandicoots (Marsupialia; Peramelemorphia) from the late Oligocene of Australia. *Palaeontologia Electronica* **16**, 13A.1–13A.52. doi:10.26879/363

394. Travouillon KJ, Cooke BN, Archer M, Hand SJ (2014) Revision of basal macropodids from the Riversleigh World Heritage Area with descriptions of new material of *Ganguroo bilamina* Cooke, 1997 and a new species. *Palaeontologica Electronica* **17**, 20A.1–20A.34. doi:10.26879/402

395. Turner S (1982) *Saurichthys* (Pisces, Actinopterygii) from the Early Triassic of Queensland. *Memoirs of the Queensland Museum* **20**, 545–551.

396. Turner S, Bean LB, Dettman M, McKellar JL, McLoughlin S, Thulborn T (2009) Australian Jurassic sedimentary and fossil successions: current work and future prospects for marine and non-marine correlation. *GFF* **131**, 49–70 [Geologiska Föreningen i Stockholm Förhandlingar]. doi:10.1080/11035890902924877

397. Turner S, Burrow CJ, Warren A (2005) *Gyracanthides hawkinsi* gen. et sp. nov (Acanthodii: Gyracanthidae) from the Lower Carboniferous of Queensland with a review of gyracanthid taxa. *Palaeontology* **48**, 963–1006. doi:10.1111/j.1475-4983.2005.00479.x

398. Turner S, Jones PJ, Draper JJ (1981) Early Devonian thelodonts (Agnatha) from the Toko Syncline, western Queensland and a review of other Australian discoveries. *Bulletin of the Bureau of Mineral Resources, Geology and Geophysics* **61**, 51–69.

399. Tyler MJ, Godthelp H (1993) A new species of *Lechriodus* Boulenger (Anura: Leptodactylidae) from the early Eocene of Queensland. *Transactions of the Royal Society of South Australia* **117**, 187–189.

400. Vajda V, Raine JI (2010) A palynological investigation of plesiosaur-bearing rocks from the Upper Cretaceous Tahora Formation, Mangahouanga, New Zealand. *Alcheringa* **34**, 359–374. doi:10.1080/03115518.2010.486642

401. Van Dyck S (1981) *Antechinus puteus* (Marsupialia: Dasyuridae), a new fossil species from the Texas Caves, southeastern Queensland. *Australian Mammalogy* **5**, 59–68.

402. Wade RT (1941) The Jurassic fishes of New South Wales. *Journal and Proceedings of the Royal Society of New South Wales* **75**, 71–84.

403. Waldman M (1971) *Fish from the Freshwater Lower Cretaceous of Victoria, Australia, with Comments on the Palaeo-environment*. Palaeontological Association, London.

404. Warren AA (1980) *Parotosuchus* from the Early Triassic of Queensland and Western Australia. *Alcheringa* **4**, 25–36. doi:10.1080/03115518008558978

405. Warren AA (2007) New data on *Ossinodus pueri*, a stem tetrapod from the Early Carboniferous of Australia. *Journal of Vertebrate Paleontology* **27**, 850–862. doi:10.1671/0272-4634(2007)27[850:NDOOPA]2.0.CO;2

406. Warren AA, Hutchinson MN (1988) A new capitosaurid amphibian from the Early Triassic of Queensland, and the ontogeny of the capitosaur skull. *Palaeontology* **31**, 857–876.

407. Warren AA, Marsicano C (1998) Revision of the Brachyopidae (Temnospondyli) from the Triassic of the Sydney, Carnarvon and Tasmania Basins, Australia. *Alcheringa* **22**, 329–342. doi:10.1080/03115519808619331

408. Warren AA, Marsicano C (2000) *Banksiops*, a replacement name for *Banksia townrowi* (Amphibia, Temnospondyli). *Journal of Vertebrate Paleontology* **20**, 186. doi:10.1671/0272-4634(2000)020[0186:BARNFB]2.0.CO;2

409. Warren AA, Rich PV, Rich TH (1997) The last, last labyrinthodonts? *Palaeontographica. Abteilung A, Paläozoologie, Stratigraphie* **247**, 1–24.

410. Warren AA, Turner S (2004) The first stem tetrapod from the Lower Carboniferous of Gondwana. *Palaeontology* **47**, 151–184. doi:10.1111/j.0031-0239.2004.00353.x

411. Weber E, Hesse A (1995) The systematic position of *Aptornis*, a flightless bird from New Zealand. *Courier Forschunginstitut Senckenberg* **181**, 293–301.

412. Welles SP, Gregg DR (1971) Late Cretaceous marine reptiles of New Zealand. *Records of the Canterbury Museum* **9**, 1–111.

413. Wells RT, Camens AB (2018) New skeletal material sheds light on the palaeobiology of the Pleistocene marsupial carnivore, *Thylacoleo carnifex*. *PLoS One* **13**, e0208020. doi:10.1371/journal.pone.0208020

414. Wells RT, Moriarty K, Williams DLG (1984) The fossil vertebrate deposits of Victoria Fossil Cave Naracoorte: an introduction to the geology and fauna. *Australian Zoologist* **21**, 305–333.

415. Wells RT, Tedford RH (1995) *Sthenurus* (Macropodidae: Marsupialia) from the Pleistocene of Lake Callabonna, South Australia. *Bulletin of the American Museum of Natural History* **225**, 1–111.

416. Westaway MC, Olley J, Grün R (2017) At least 17,000 years of coexistence: modern humans and megafauna at the Willandra Lakes, south-eastern Australia. *Quaternary Science Reviews* **157**, 206–211. doi:10.1016/j.quascirev.2016.11.031

417. White AW (2001) A new Eocene soft-shelled turtle (Trionychidae) from Murgon, south-eastern Queensland. *Memoirs of the Association of Australasian Palaeontologists* **25**, 37–43.

418. White AW, Archer M (1994) *Emydura lavarackorum*, a new Pleistocene turtle (Pleurodira: Chelidae) from fluviatile deposits at Riversleigh, northwestern Queensland. *Records of the South Australian Museum* **27**, 159–167.

419. White AW, Worthy TH, Hawkins S, Bedford S, Spriggs M (2010) Megafaunal meiolaniid horned turtles survived until early human settlement in Vanuatu, southwest Pacific. *Proceedings of the National Academy of Sciences of the United States of America* **107**, 15512–15516. doi:10.1073/pnas.1005780107

420. White MA, Cook AG, Hocknull SA, Sloan T, Sinapius GHK, Elliott DA (2012) New forearm elements discovered of holotype specimen *Australovenator wintonensis* from Winton, Queensland, Australia. *PLoS One* **7**, e39364. doi:10.1371/journal.pone.0039364

421. White MA, Cook AG, Klinkhamer AJ, Elliott DA (2016) The pes of *Australovenator wintonensis* (Theropoda: Megaraptoridae): analysis of the pedal range of motion and biological restoration. *PeerJ* **4**, e2312. doi:10.7717/peerj.2312

422. White MA, Falkingham PL, Cook AG, Hocknull SA, Elliott DA (2013) Morphological comparisons of metacarpal I for *Australovenator wintonensis* and *Rapator ornitholestoides*: implications for their taxonomic relationships. *Alcheringa* **37**, 435–441. doi:10.1080/03115518.2013.770221

423. Wiffen J (1980) *Moanasaurus*, a new genus of marine reptile (Family Mosasauridae) from the Upper Cretaceous of North Island. *New Zealand Journal of Geology and Geophysics* **23**, 507–528. doi:10.1080/00288306.1980.10424122

424. Wiffen J (1990) New mosasaurs (Reptilia; Family Mosasauridae) from the Upper Cretaceous of North Island. *New Zealand Journal of Geology and Geophysics* **33**, 67–85. doi:10.1080/00288306.1990.10427574

425. Wiffen J, Moisley WL (1986) Late Cretaceous reptiles (Families Elasmosauridae and Pliosauridae) from the Mangahouanga Stream, North Island. *New Zealand Journal of Geology and Geophysics* **29**, 205–252. doi:10.1080/00288306.1986.10427535

426. Wiffen J, Molnar RE (1988) First pterosaur from New Zealand. *Alcheringa* **12**, 53–59. doi:10.1080/03115518808618996

427. Willis PMA (1993) *Trilophosuchus rackhami*, gen. et sp. nov., a new crocodilian from the early Miocene Limestones of Riversleigh, northwestern Queensland. *Journal of Vertebrate Paleontology* **13**, 90–98. doi:10.1080/02724634.1993.10011489

428. Willis PMA (1997) New crocodilians from the late Oligocene White Hunter Site, Riversleigh, northwestern Queensland. *Memoirs of the Queensland Museum* **41**, 423–438.

429. Willis PMA (1997) Review of fossil crocodilians from Australasia. *Australian Zoologist* **30**, 287–298. doi:10.7882/AZ.1997.004

430. Willis PMA (2001) New crocodilian material from the Miocene of Riversleigh (northwestern Queensland, Australia). In *Crocodilian Biology and Evolution*. (Eds GC Grigg, F Seebacher and CE Franklin) pp. 64–74. Surrey Beatty & Sons, Sydney.

431. Willis PMA, Molnar RE (1991) A new Middle Tertiary crocodile from Lake Palankarinna, South Australia. *Records of the South Australian Museum* **25**, 39–55.

432. Willis PMA, Molnar RE (1997) A Review of the Plio-Pleistocene crocodilian genus *Pallimnarchus*. *Proceedings of the Linnean Society of New South Wales* **117**, 223–242.

433. Willis PMA, Molnar RE, Scanlon JD (1993) An early Eocene crocodylian from Murgon, southeastern Queensland. *Kaupia Darmstädter Beiträge zur Naturgeschichte* **3**, 25–32.

434. Wood JR, Wilmshurst JM, Richardson SJ, Rawlence NJ, Wagstaff SJ, Worthy TH, Cooper A (2013) Resolving lost herbivore community structure using coprolites of four sympatric moa species (Aves: Dinornithiformes). *Proceedings of the National Academy of Sciences of the United States of America* **110**, 16910–16915. doi:10.1073/pnas.1307700110

435. Woodburne MO (1967) The Alcoota Fauna, central Australia. *Bulletin of the Bureau of Mineral Resources, Geology and Geophysics* **87**, 1–187.

436. Woodburne MO, Clemens WA (Eds) (1986) *Revision of the Ektopodontidae (Mammalia; Marsupialia; Phalangeroidea) of the Australian Neogene.* UC Publications in Geological Sciences, Berkeley.

437. Woodburne MO, Macfadden BJ, Case JA, Springer MS, Pledge NS, Power JD, et al. (1994) Land mammal biostratigraphy and magnetostratigraphy of the Etadunna Formation (Late Oligocene) of South Australia. *Journal of Vertebrate Paleontology* **13**, 483–515. doi:10.1080/02724634.1994.10011527

438. Woodhead J, Hand SJ, Archer M, Graham I, Sniderman K, Arena DA, Black KH, Godthelp H, Creaser P, Price E (2016) Developing a radiometrically-dated chronologic sequence for Neogene biotic change in Australia, from the Riversleigh World Heritage Area of Queensland. *Gondwana Research* **29**, 153–167. doi:10.1016/j.gr.2014.10.004

439. Woodward AS (1888) Notes on the extinct reptilian genera *Megalania*, Owen, and *Meiolania*, Owen. *Annals and Magazine of Natural History. Series 6* **1**(2), 85–89. doi:10.1080/00222938809460686

440. Woodward AS (1895) The fossil fishes of the Talbragar Beds (Jurassic). *Palaeontology* **9**, 1–27.

441. Woodward AS (1906) On a Carboniferous fish fauna from the Mansfield district, Victoria. *Memoirs of the National Museum, Melbourne* **1**, 1–32. doi:10.24199/j.mmv.1906.1.01

442. Worthy TH (1997) The Quaternary fossil fauna of South Canterbury, South Island, New Zealand. *Journal of the Royal Society of New Zealand* **27**, 67–162. doi:10.1080/03014223.1997.9517528

443. Worthy TH (2000) The fossil megapodes (Aves: Megapodiidae) of Fiji with descriptions of a new genus and two new species. *Journal of the Royal Society of New Zealand* **30**, 337–364. doi:10.1080/03014223.2000.9517627

444. Worthy TH (2001) A giant flightless pigeon gen. et sp. nov. and a new species of *Ducula* (Aves: Columbidae), from Quaternary deposits in Fiji. *Journal of the Royal Society of New Zealand* **31**, 763–794. doi:10.1080/03014223.2001.9517673

445. Worthy TH (2008) Pliocene waterfowl (Aves: Anseriformes) from South Australia and a new genus and species. *Emu - Austral Ornithology* **108**, 153–165. doi:10.1071/MU07063

446. Worthy TH, Burley D (2020) The prehistoric avifaunas from the Kingdom of Tonga. *Zoological Journal of the Linnean Society* **189**, 998–1045. doi:10.1093/zoolinnean/zlz110

447. Worthy TH, Anderson AJ, Molnar RE (1999) Megafaunal expression in a land without mammals: the first fossil faunas from terrestrial deposits in Fiji. *Senckenbergiana Biologica* **79**, 237–242.

448. Worthy TH, Boles WE (2011) *Australlus*, a new genus for *Gallinula disneyi* (Aves: Rallidae) and a description of a new species from Oligo-Miocene deposits at Riversleigh, northwestern Queensland, Australia. *Records of the Australian Museum* **63**, 61–77. doi:10.3853/j.0067-1975.63.2011.1563

449. Worthy TH, Hand SJ, Archer M (2014) Phylogenetic relationships of the Australian Oligo-Miocene ratite *Emuarius gidju* Casuariidae. *Integrative Zoology* **9**, 148–166. doi:10.1111/1749-4877.12050

450. Worthy TH, Hand SJ, Archer M, Scofield RP, de Pietri VL (2019) Evidence for a giant parrot from the Early Miocene of New Zealand. *Biology Letters* **15**(8), 20190467. doi:10.1098/rsbl.2019.0467

451. Worthy TH, Hand SJ, Nguyen JMT, Tennyson AJD, Worthy JP, Scofield RP, Boles WE, Archer M (2010) Biogeographical and phylogenetic implications

of an Early Miocene wren (Aves: Passeriformes: Acanthisittidae) from New Zealand. *Journal of Vertebrate Paleontology* **30**, 479–498. doi:10.1080/02724631003618033

452. Worthy TH, Holdaway RN (2002) *The Lost World of the Moa: Prehistoric Life of New Zealand.* Indiana University Press, Bloomington.

453. Worthy TH, Holdaway RN, Sorenson MD, Cooper AC (1997) Description of the first complete skeleton of the extinct New Zealand goose *Cnemiornis calcitrans* Owen, (Aves: Anatidae), and a reassessment of the relationships of *Cnemiornis. Journal of Zoology* **243**, 695–718. doi:10.1111/j.1469-7998.1997.tb01971.x

454. Worthy TH, Lee MSY (2008) Affinities of Miocene waterfowl (Anatidae: *Manuherikia, Dunstanetta* and *Miotadorna*) from the St Bathans Fauna, New Zealand. *Palaeontology* **51**, 677–708. doi:10.1111/j.1475-4983.2008.00778.x

455. Worthy TH, de Pietri VL, Scofield RP (2017) Recent advances in avian palaeobiology in New Zealand with implications for understanding New Zealand's geological, climatic and evolutionary histories. *New Zealand Journal of Zoology* **44**, 177–211. doi:10.1080/03014223.2017.1307235

456. Worthy TH, Scofield RP (2012) Twenty-first century advances in knowledge of the biology of moa (Aves: Dinornithiformes): a new morphological analysis and diagnoses revised. *New Zealand Journal of Zoology* **39**, 87–153. doi:10.1080/03014223.2012.665060

457. Worthy TH, Tennyson AJD, Archer M, Musser AM, Hand SJ, Jones C, Douglas BJ, McNamara JA, Beck RMD (2006) Miocene mammal reveals a Mesozoic ghost lineage on insular New Zealand, southwest Pacific. *Proceedings of the National Academy of Sciences of the United States of America* **103**, 19419–19423. doi:10.1073/pnas.0605684103

458. Worthy TH, Tennyson AJD, Hand SJ, Scofield RP (2008) A new species of the diving duck *Manuherikia* and evidence for geese (Aves: Anatidae: Anserinae) in the St Bathans Fauna (Early Miocene), New Zealand. *Journal of the Royal Society of New Zealand* **38**, 97–114. doi:10.1080/03014220809510549

459. Worthy TH, Tennyson AJD, Jones C, McNamara JA, Douglas BJ (2007) Miocene waterfowl and other birds from Central Otago, New Zealand. *Journal of Systematic Palaeontology* **5**, 1–39. doi:10.1017/S1477201906001957

460. Worthy TH, Tennyson AJD, Scofield RP (2011) An Early Miocene diversity of parrots (Aves, Strigopidae, Nestorinae) from New Zealand. *Journal of Vertebrate Paleontology* **31**, 1102–1116. doi:10.1080/02724634.2011.595857

461. Worthy TH, Worthy JP, Tennyson AJD, Salisbury SW, Hand SJ, Scofield RP (2013) Miocene fossils show that kiwi (*Apteryx*, Apterygidae) are probably not phyletic dwarves. In *Proceedings of the 8th International Meeting of the Society of Avian Paleontology and Evolution.* (Eds UB Göhlich and A Kroh) pp. 63–80. Natural History Museum, Vienna.

462. Wroe S (1998) A new bone-cracking dasyurid from the Miocene of Riversleigh, northwestern Queensland. *Alcheringa* **22**, 277–284. doi:10.1080/03115519808619205

463. Wroe S, Crowther M, Dortch J, Chong J (2004) The size of the largest marsupial and why it matters. *Proceedings. Biological Sciences* **271**, S34–S36.

464. Wroe S, Field JH, Archer M, Grayson DK, Price GJ, Louys J, Faith JT, Webb GE, Davidson I, Mooney SD (2013) Climate change frames debate over the extinction of megafauna in Sahul (Pleistocene Australia-New Guinea). *Proceedings of the National Academy of Sciences of the United States of America* **110**, 8777–8781. doi:10.1073/pnas.1302698110

465. Yates AM (2015) New craniodental remains of *Wakaleo alcootaensis* (Diprotodontia: Thylacoleonidae) a carnivorous marsupial from the late Miocene Alcoota Local Fauna of the Northern Territory, Australia. *PeerJ* **3**, e1408. doi:10.7717/peerj.1408

466. Young GC (1991) The first armoured agnathan vertebrates from the Devonian of Australia. In *Early Vertebrates and Related Problems of Evolutionary Biology.* (Eds MM Chang, YH Liu and GR Zhang) pp. 67–85. Science Press, Beijing.

467. Young GC (2004) Large brachythoracid arthrodires (placoderm fishes) from the early Devonian of Wee Jasper, New South Wales, Australia, with a discussion of basal brachythoracid characters. *Journal of Vertebrate Paleontology* **24**, 1–17. doi:10.1671/1942-1

468. Young GC (2005) Early Devonian arthrodire remains (Placodermi? Holonematidae) from the Burrinjuck area, New South Wales, Australia. *Geodiversitas* **27**, 201–219.

469. Young GC (2011) Wee Jasper–Lake Burrinjuck fossil fish sites: scientific background to national heritage nomination. *Proceedings of the Linnean Society of New South Wales* **132**, 83–107

470. Young GC, Goujet D (2003) Devonian fish remains from the Dulcie Sandstone and Cravens Peak beds, Georgina basin, central Australia. *Records of the Western Australian Museum* **65**(supp.), 1–85. doi:10.18195/issn.0313-122x.65.2003.001-085

471. Zuccon D, Ericson PGP (2012) Molecular and morphological evidences place the extinct New Zealand endemic *Turnagra capensis* in the Oriolidae. *Molecular Phylogenetics and Evolution* **62**, 414–426. doi:10.1016/j.ympev.2011.10.013

Index